THE FEMALE BRAIN

Louann Brizendine, M.D.

BANTAM BOOKS

LONDON • TORONTO • SYDNEY • AUCKLAND • JOHANNESBURG

The information in this book has been compiled by way of general guidance in relation to the specific subjects addressed, but is not a substitute for medical advice. Always consult a qualified medical practitioner before starting, changing or stopping any medical treatment. So far as the author is aware, the information given is correct and up to date as at September 2006. Practice, laws and regulations all change, and the reader should obtain up-to-date professional advice on any such issues. The author and publishers disclaim, as far as the law allows, any liability arising directly or indirectly from the use, or misuse, of the information contained in this book.

TRANSWORLD PUBLISHERS
61–63 Uxbridge Road, London W5 5SA
A Random House Group Company
www.booksattransworld.co.uk

THE FEMALE BRAIN
A BANTAM BOOK: 9780553818499

First published in Great Britain
in 2007 by Bantam Press
a division of Transworld Publishers
Bantam edition published 2007

A CIP catalogue record for this book
is available from the British Library.

Addresses for Random House Group Ltd companies outside the UK
can be found at: www.randomhouse.co.uk
The Random House Group Ltd Reg. No. 954009

The Random House Group Limited supports The Forest Stewardship
Council (FSC), the leading international forest certification organisation.
All our titles that are printed on Greenpeace approved FSC certified paper
carry the FSC logo. Our paper procurement policy can be found at:
www.rbooks.co.uk/environment.

Typeset in 10.5/14pt Bell MT by
Falcon Oast Graphic Art Ltd.

Printed in the UK by
CPI Cox & Wyman, Reading, RG1 8EX

4 6 8 10 9 7 5

Praise for Dr Louann Brizendine's *The Female Brain*

'Finally, a satisfying answer to Freud's question, "What does a woman want?" Louann Brizendine has done a great favour for every man who wants to understand the puzzling women in his life. A breezy and enlightening guide to women – and a must-read for men' **Daniel Goleman, author of** *Social Intelligence*

'The author's greatest gift to her readers is the way she takes us through the stages of a woman's life to show the influence of hormone levels on every decision. It's not just a matter of biology [but] also of how biology affects perception and our ability to function' *Los Angeles Times*

'Readers will sift through the case studies, research findings and scientific conjectures gathered in this non-technical book and be intrigued . . .' **Deborah Tannen,** *Washington Post*

'This comprehensive new look at the hormonal roller coaster that rules women's lives down to the cellular level offers a trove of information, as well as some stunning insights . . . Fully accessible, [it] provides a fascinating look at the life cycle of the female brain . . .' *Publishers Weekly*

'Brizendine's conclusions will seem like common sense to some and nothing short of heresy to others . . . She not only discusses the biological reasons girls gravitate to dolls instead of trucks but tracks the hormonal imperatives at play when a teenage female becomes obsessed with text messaging and shopping. She also explains how changing brain chemistry can prompt a post-menopausal woman to forgo marriage counselling and dial up a divorce lawyer instead' *Newsweek*

www.booksattransworld.co.uk

'An eye-opening account of the biological foundations of human behaviour. Destined to become a classic . . .' **Marilyn Yalom, author of** *A History of the Wife*

'Brizendine is well qualified to explain how changing hormone levels influence behavior through adolescence, pregnancy, motherhood, and menopause. Here, she writes of her research, shares anecdotes from her clinical practice, and examines the latest scientific studies.' *Library Journal*

'A timely, insightful, readable and an altogether magnificent book.' **Sarah Blaffer Hrdy, author of** *Mother Nature*

'Do men and women think alike? Well, that's a no-brainer. Of course not. Although 99% of our genes are the same, that 1% makes all the difference, as neurologist Louann Brizendine explains in her elegant new book.' *Oprah Magazine*

For my husband,
Samuel Barondes,

My son,
John Whitney Brizendine,

And in loving memory of
Louise Ann Brizendine

CONTENTS

ACKNOWLEDGMENTS

THIS BOOK HAD its beginnings during my educational years at the University of California, Berkeley; Yale; Harvard; and University College, London, so I would like to thank the teachers and fellow students who most influenced my thinking during those years: Frank Beach, Mina Bissel, Henry Black, Bill Bynum, Dennis Charney, Marion Diamond, Marilyn Farquar, Carol Gilligan, Paul Greengard, Tom Guteil, Les Havens, Florence Haseltine, Marjorie Hayes, Peter Hornick, Stanley Jackson, Valerie Jacoby, Kathleen Kells, Kathy Kelly, Adrienne Larkin, Howard Levitin, Mel Lewis, Charlotte McKenzie, David Mann, Daniel Mazia, William Meissner, Jonathan Muller, Fred Naftolin, George Palade, Roy Porter, Sherry Ryan, Carl Salzman, Leon Shapiro, Rick Shelton, Gunter Stent, Frank Thomas, Janet Thompson, George Vaillant, Roger Wallace, Clyde Willson, Fred Wilt, and Richard Wollheim.

During my years on the faculty at Harvard and the University of California, San Francisco, my thinking has been influenced by Bruce Ames, Cori Bargmann, Regina Casper, Francis Crick, Mary Dallman, Herb Goldings, Deborah Grady, Joel Kramer, Fernand Labrie, Jeanne Leventhal, Sindy Mellon, Michael Merzenich, Joseph Morales, Eugene Roberts, Laurel Samuels, Carla Shatz, Stephen Stahl, Elaine Storm, Marc Tessier-Lavigne, Rebecca Turner, Victor Viau, Owen Wolkowitz, and Chuck Yingling.

My colleagues, staff, residents, medical students, and patients in the Women's and Teen Girls' Mood and Hormone Clinic have contributed in many ways to this work: Denise Albert, Raya Almufti, Amy Berlin, Cathy Christensen, Karen Cliffe, Allison Doupe, Judy Eastwood, Louise Forrest, Adrienne Fratini, Lyn Gracie, Marcie Hall-Mennes, Steve Hamilton, Caitlin Hasser, Dannah Hirsch, Susie Hobbins, Fatima Imara, Lori Lavinthal, Karen Leo, Shana Levy, Katherine Malouh, Faina Nosolovo, Sarah Prolifet, Jeanne St. Pierre, Veronica Saleh, Sharon Smart, Alla Spivak, Elizabeth Springer, Claire Wilcox, and Emily Wood.

I also thank my other colleagues, students, and staff at Langley Porter Psychiatric Institute and UCSF whose contributions I have appreciated: Alison Adcock, Regina Armas, Jim Asp, Renee Binder, Kathryn Bishop, Mike Bishop, Alla Borik, Carol Brodsky, Marie Caffey, Lin Cerles, Robin Cooper, Haile Debas, Andrea DiRocchi, Glenn Elliott, Stu Eisendrath, Leon Epstein, Laura Esserman, Ellen Haller, Dixie Horning, Marc Jacobs, Nancy Kaltreider, David Kessler, Michael Kirsch, Laurel Koepernick, Rick Lannon, Bev Lehr, Descartes Li, Jonathan Lichtmacher, Elaine Lonnergan, Alan Louie, Theresa McGinness, Robert Malenka, Charlie Marmar, Miriam Martinez, Craig Nelson, Kim Norman, Chad Peterson, Anne Poirier, Astrid Prackatzch, Victor Reus, John Rubenstein, Bryna Segal, Lynn Shroeder, John Sikorski, Susan Smiga, Anna Spielvogel, David Taylor, Larry Tecott, Renee Valdez, Craig Van Dyke, Mark Van Zastrow, Susan Voglmaier, John Young, and Leonard Zegans.

I am very grateful to those who have read and critiqued drafts of this book: Carolyn Balkenhol, Marcia Barinaga, Elizabeth Barondes, Diana Brizendine, Sue Carter, Sarah Cheyette, Diane Cirrincione, Theresa Crivello, Jennifer Cummings, Pat Dodson,

Janet Durant, Jay Giedd, Mel Grumbach, Dannah Hirsch, Sarah Hrdy, Cynthia Kenyon, Adrienne Larkin, Jude Lange, Jim Leckman, Louisa Llanes, Rachel Llanes, Eleanor Maccoby, Judith Martin, Diane Middlebrook, Nancy Milliken, Cathy Olney, Linda Pastan, Liz Perle, Lisa Queen, Rachel Rokicki, Dana Slatkin, Millicent Tomkins, and Myrna Weissman.

The work presented here has particularly benefited from the research, writings, and advice of Marty Altemus, Arthur Aron, Simon Baron-Cohen, Jill Becker, Andreas Bartels, Lucy Brown, David Buss, Larry Cahill, Anne Campbell, Sue Carter, Lee Cohen, Susan Davis, Helen Fisher, Jay Giedd, Jill Goldstein, Mel Grumbach, Andy Guay, Melissa Hines, Nancy Hopkins, Sarah Hrdy, Tom Insel, Bob Jaffe, Martha McClintock, Erin McClure, Eleanor Maccoby, Bruce McEwen, Michael Meaney, Barbara Parry, Don Pfaff, Cathy Roca, David Rubinow, Robert Sapolsky, Peter Schmidt, Nirao Shah, Barbara Sherwin, Elizabeth Spelke, Shelley Taylor, Kristin Uvnäs-Moberg, Sandra Witelson, Sam Yen, Kimberly Yonkers, and Elizabeth Young.

I also thank supporters with whom I have had lively and influential conversations over the past few years about the female brain: Bruce Ames, Giovanna Ames, Elizabeth Barondes, Jessica Barondes, Lynne Krilich Benioff, Marc Benioff, ReVeta Bowers, Larry Ellison, Melanie Craft Ellison, Cathy Fink, Steve Fink, Milton Friedman, Hope Frye, Donna Furth, Alan Goldberg, Andy Grove, Eva Grove, Anne Hoops, Jerry Jampolsky, Laurene Powell Jobs, Tom Kornberg, Josh Lederberg, Marguerite Lederberg, Deborah Leff, Sharon Agopian Melodia, Shannon O'Rourke, Judy Rapoport, Jeanne Robertson, Sandy Robertson, Joan Ryan, Dagmar Searle, John Searle, Garen Staglin, Shari Staglin, Millicent Tomkins, Jim Watson, Meredith

White, Barbara Willenborg, Marilyn Yalom, and Jody Kornberg Yeary.

I would also like to thank the individuals and private foundations that have supported my work: Lynne and Marc Benioff, the Lawrence Ellison Medical Foundation, National Center for Excellence in Women's Health at UCSF, the Osher Foundation, the Salesforce.com Foundation, the Staglin Family Music Festival for Mental Health, the Stanley Foundation, and the UCSF Department of Psychiatry.

This book was initially developed through the skill and talent of Susan Wels, who helped me write the first draft and organize vast amounts of material. I owe her the greatest debt of gratitude.

I am very thankful to Liz Perle, who first persuaded me to write this book, and to the others who believed in it and worked hard to make it happen: Susan Brown, Rachel Lehmann-Haupt, Deborah Chiel, Marc Haeringer, and Rachel Rokicki. My agent, Lisa Queen of Queen Literary, has been a terrific supporter and has made many brilliant suggestions throughout this process.

I am especially grateful to Amy Hertz, vice president and publisher of Morgan Road Books, who had a vision for this project from the beginning and kept demanding excellence and crafting revisions to create a narrative in which the science comes alive.

I also want to thank my son, Whitney, who tolerated this long and demanding project with grace and made important contributions to the teen chapter.

Most of all I thank my husband and soul mate, Sam Barondes, for his wisdom, endless patience, editorial advice, scientific insight, love, and support.

THE FEMALE BRAIN

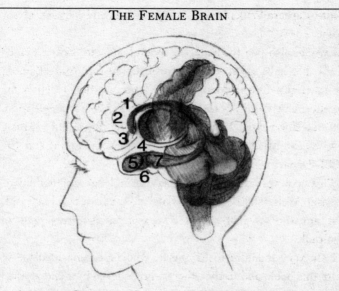

1. ANTERIOR CINGULATE CORTEX (ACC): Weighs options, makes decisions. It's the worry-wort center, and it's larger in women than in men.

2. PREFRONTAL CORTEX (PFC): The queen that rules the emotions and keeps them from going wild. It puts the brakes on the amygdala. Larger in women, and matures faster in women than in men by one to two years.

3. INSULA: The center that processes gut feelings. Larger and more active in women.

4. HYPOTHALAMUS: The conductor of the hormonal symphony; kicks the gonads into gear. Starts pumping earlier in life in women.

5. AMYGDALA: The wild beast within; the instinctual core, tamed only by the PFC. Larger in men.

6. PITUITARY GLAND: Produces hormones of fertility, milk production, and nurturing behaviour. Helps turn on the mommy brain.

7. HIPPOCAMPUS: The elephant that never forgets a fight, a romantic encounter, or a tender moment – and won't let you forget it, either. Larger and more active in women.

(in other words, how hormones affect a woman's brain)

THE ONES YOUR doctor knows about

ESTROGEN—the queen: powerful, in control, all-consuming; sometimes all business, sometimes an aggressive seductress; friend of dopamine, serotonin, oxytocin, acetylcholine, and norepinephrine (the feel-good brain chemicals).

PROGESTERONE—in the background but a powerful sister to estrogen; intermittently appears and sometimes is a storm cloud reversing the effects of estrogen; other times is a mellowing agent; mother of allopregnenolone (the brain's Valium, i.e., chill pill).

TESTOSTERONE—fast, assertive, focused, all-consuming, masculine; forceful seducer; aggressive, unfeeling; has no time for cuddling.

THE ONES YOUR doctor may not know about that also affect a woman's brain

OXYTOCIN—fluffy, purring kitty; cuddly, nurturing, earth mother; the good witch Glinda in *The Wizard of Oz*; finds pleasure in helping and serving; sister to vasopressin (the male socializing hormone), sister to estrogen, friend of dopamine (another feel-good brain chemical).

CORTISOL—frizzled, frazzled, stressed out; highly sensitive, physically and emotionally.

VASOPRESSIN—secretive, in the background, subtle aggressive male energies; brother to testosterone, brother to oxytocin (makes you want to connect in an active, male way, as does oxytocin).

DHEA—reservoir of all the hormones; omnipresent, pervasive, sustaining mist of life; energizing; father and mother of testosterone and estrogen, nicknamed "the mother hormone," the Zeus and Hera of hormones; robustly present in youth, wanes to nothing in old age.

ANDROSTENEDIONE—the mother of testosterone in the ovaries; supply of sassiness; high-spirited in youth, wanes at menopause, dies with the ovaries.

ALLOPREGNENOLONE—the luxurious, soothing, mellowing daughter of progesterone; without her, we are crabby; she is sedating, calming, easing; neutralizes any stress, but as soon as she leaves, all is irritable withdrawal; her sudden departure is the central story of PMS, the three or four days before a woman's period starts.

PHASES OF
A FEMALE'S LIFE

HORMONES CAN DETERMINE what the brain is interested in doing. They help guide nurturing, social, sexual, and aggressive behaviors. They can affect being talkative, being flirtatious, giving or attending parties, writing thank-you notes, planning children's play dates, cuddling, grooming, worrying about hurting the feelings of others, being competitive, masturbating and initiating sex.

PHASES OF A FEMALE'S LIFE

	MAJOR HORMONE CHANGES	WHAT FEMALES HAVE THAT MALES DON'T
FETAL	Brain growth and development left unperturbed by the high testosterone of a male brain	Brain cells are XX, which means more genes for fast brain development and female-specific circuits
GIRLHOOD	Estrogen is secreted in massive amounts from age 6 to 24 months, then the juvenile pause turns off hormones	High estrogen for up to 2 years after birth
PUBERTY	Estrogen, progesterone, and testosterone increase and begin to cycle monthly	More estrogen and less testosterone; girls' brains develop 2 years earlier than boys'
SEXUAL MATURITY, SINGLE WOMAN	Estrogen, progesterone, and testosterone change every day of the month	More focus on relationships, finding a lifelong mate, and choosing a career or job compatible with raising a family
PRENANCY	Huge increase in progesterone, estrogen	Focus more on nesting, how the family will be provided for; less on career competition
BREAST FEEDING	Oxytocin, prolactin	Focus more exclusively on the baby
CHILD REARING	Oxytocin; cycling estrogen, progesterone, and testosterone	Less interest in sex, more worry about kids
PERIMENOPAUSE	Erratically cycling estrogen, progesterone, and testosterone	Fluctuating interest in sex, erratic sleep, more fatigue, worry, moods, hot flashes and irritability
MENOPAUSE	Low estrogen and no progresterone, high FSH/LH	The last precipitous brain change caused by hormones
POSTMENOPAUSE	Low, steady estrogen and testosterone, lower oxytocin	More calmness

FEMALE-SPECIFIC BRAIN CHANGES	REALITY CHANGE
Female brain circuits for communication, gut feelings, emotional memory, and anger suppression grow unabated – there is no high testosterone of the male around to kill all those cells.	More brain circuits for communication, reading emotions, social nuance, nurturing skills, able to use both sides of the brain
Verbal and emotional circuits are enhanced	Major interest in playing and having fun in connection with other girls, not boys
Increased sensitivity and growth of stress, verbal, emotion, and sex circuits	Major interest in sexual attractiveness, desperate love interests, avoidance of parents
Earlier maturation of decision-making and emotional control circuits	Major interests in finding a mate, love, career development
Stress circuits suppressed, brain clamed by progesterone; brain shrinks hormones from the fetus and placenta take over brain and body	Major interest in own physical well-being, coping with fatigue, nausea, and hunger, and not damaging the fetus; surviving in the workplace; and planning maternity leave
Stress circuits still suppressed; sex amd emotion circuits hijacked by infant care	Major focus on coping with fatigue, sore nipples, breast milk production, making it through the next 24 hours
Increased function of stress, worry, and emotional bonding circuits	Major interest in well-being, development, education, and safety of kids, coping with increased stress and work
Decreasing sensitivity to estrogen in certain brain circuits	Major interest in surviving day to day and coping with the physical and emotional ups and downs
Circuits fueled by estorgen, oxytocin, and progesterone decline	Major interest in staying healthy, improving well-being and embracing new challenges
Circuits less reactive to stress, less emotional	Major interest in doing what *you* want to do, less interest in taking care of others

THE FEMALE BRAIN

What Makes Us Women

MORE THAN 99 percent of male and female genetic coding is exactly the same. Out of the thirty thousand genes in the human genome, the less than one percent variation between the sexes is small. But that percentage difference influences every single cell in our bodies—from the nerves that register pleasure and pain to the neurons that transmit perception, thoughts, feelings, and emotions.

To the observing eye, the brains of females and males are not the same. Male brains are larger by about 9 percent, even after correcting for body size. In the nineteenth century, scientists took this to mean that women had less mental capacity than men. Women and men, however, have the same number of brain cells. The cells are just packed more densely in women—cinched corsetlike into a smaller skull.

For much of the twentieth century, most scientists assumed that women were essentially small men, neurologically and in every other sense except for their reproductive functions. That assumption has been at the heart of enduring misunderstandings about female psychology and physiology. When you look a little deeper into the brain differences, they reveal what makes women women and men men.

Until the 1990s, researchers paid little attention to female physiology, neuroanatomy, or psychology separate from that of men. I saw this oversight firsthand during my undergraduate years in neurobiology at Berkeley in the 1970s, during my medical education at Yale, and during my training in psychiatry at the Massachusetts Mental Health Center at Harvard Medical School. While enrolled at each of these institutions, I learned little or nothing about female biological or neurological difference outside of pregnancy. When a professor presented a study about animal behavior one day at Yale, I raised my hand and asked what the research findings were for females in that study. The male professor dismissed my question, stating, "We never use females in these studies—their menstrual cycles would just mess up the data."

The little research that was available, however, suggested that the brain differences, though subtle, were profound. As a resident in psychiatry, I became fascinated by the fact that there was a two-to-one ratio of depression in women compared with men. No one was offering any clear reasons for this discrepancy. Because I had gone to college at the peak of the feminist movement, my personal explanations ran toward the political and the psychological. I took the typical 1970s stance that the patriarchy of Western culture must have been the culprit. It must have kept women down and made them less functional than men. But that explanation alone didn't seem to fit: new studies were uncovering the same depression ratio worldwide. I started to think that something bigger, more basic and biological, was going on.

One day it struck me that male versus female depression rates didn't start to diverge until females turned twelve or thirteen—the age girls began menstruating. It appeared that the chemical changes at puberty did something in the brain to trigger more depression in women. Few scientists at the time were researching this link, and most psychiatrists, like me, had been trained in traditional psychoanalytic theory, which examined childhood experience but never considered that specific female brain chemistry might be involved. When I started taking a woman's hormonal state into account as I evaluated her psychiatrically, I discovered the massive neurological effects her hormones have during different stages of life in shaping her desires, her values, and the very way she perceives reality.

My first epiphany about the different realities created by sex hormones came when I started treating women with what I call extreme premenstrual brain syndrome. In all menstruating women, the female brain changes a little every day. Some parts of the brain change up to 25 percent every month. Things get rocky at times, but for most women, the changes are manageable. Some of my patients, though, came to me feeling so jerked around by their hormones on some days that they couldn't work or speak to anyone because they'd either burst into tears or bite someone's head off. Most weeks of the month they were engaged, intelligent, productive, and optimistic, but a mere shift in the hormonal flood to their brains on certain days left them feeling that the future looked bleak, and that they hated themselves and their lives. These thoughts felt real and solid, and these women acted on them as though they were reality and would last forever—even though they arose solely from hormonal shifts in their brains. As soon as the tides changed, they were back to their best selves. This extreme form of PMS, which is present in only a few percent of women, introduced me to how the female brain's reality can turn on a dime.

If a woman's reality could change radically from week to week, the same would have to be true of the massive hormonal changes that occur throughout a woman's life. I wanted the chance to find out more about these possibilities on a broader scale, and so, in 1994, I founded the Women's Mood and Hormone Clinic in the Department of Psychiatry at the University of California, San Francisco. It was one of the first clinics in the country dedicated to looking at women's brain states, and how neurochemistry and hormones affect their moods.

What we've found is that the female brain is so deeply affected by hormones that their influence can be said to create a woman's reality. They can shape a woman's values and desires, and tell her, day to day, what's important. Their presence is felt at every stage of life, right from birth. Each hormone state—girlhood, the adolescent years, the dating years, motherhood, and menopause—acts as fertilizer for different neurological connections that are responsible for new thoughts, emotions, and interests. Because of the fluctuations that begin as early as three months old and last until after menopause, a woman's neurological reality is not as constant as a man's. His is like a mountain that is worn away imperceptibly over the millennia by glaciers, weather, and the deep tectonic movements of the earth. Hers is more like the weather itself—constantly changing and hard to predict.

NEW BRAIN SCIENCE has rapidly transformed our view of basic neurological differences between men and women. Earlier scientists could investigate these differences only by studying the brains of cadavers or the symptoms of individuals with brain damage. But thanks to advances in genetics and noninvasive brain-imaging technology, there's been a complete revolution in neuroscientific research and theory. New tools, such as positron-emission tomography (PET) and functional magnetic resonance imaging (fMRI) scans, now allow us to see inside the human brain in real time, while it's solving problems, producing words, retrieving memories, noticing facial expressions, establishing trust, falling in love, listening to babies cry, and feeling depression, fear, and anxiety.

As a result, scientists have documented an astonishing array of structural, chemical, genetic, hormonal, and functional brain differences between women and men. We've learned that men and women have different brain sensitivities to stress and conflict. They use different brain areas and circuits to solve problems, process language, experience and store the same strong emotion. Women may remember the smallest details of their first dates, and their biggest fights, while their husbands barely remember that these things happened. Brain structure and chemistry have everything to do with why this is so.

The female and male brains process stimuli, hear, see, "sense," and gauge what others are feeling in different ways. Our distinct female and male brain operating systems are mostly compatible and adept, but they perform and accomplish the same goals and tasks using different circuits. In a German study, researchers conducted brain scans of men and women while they mentally rotated abstract, three-dimensional shapes. There were no performance differences between the men and women, but there were significant, sex-specific differences in the brain circuits they activated to complete the task. Women triggered brain pathways linked to visual identification and spent more time than men picturing the objects in their minds. This fact merely meant that it took women longer to get to the same answer. It also showed that females perform all the cognitive functions males perform—they just do so by using different brain circuits.

Under a microscope or an fMRI scan, the differences between male and female brains are revealed to be complex and widespread. In the brain centers for language and hearing, for example, women have 11 percent more neurons than men. The principal hub of both emotion and memory formation—the hippocampus—is also larger in the female brain, as is the brain circuitry for language and observing emotions in others. This means that women are, on average, better at expressing emotions and remembering the details of emotional events. Men, by contrast, have two and a half times the brain space devoted to sexual drive as well as larger brain centers for action and aggression. Sexual thoughts float through a man's brain many times each day on average, and through a woman's only once a day. Perhaps three to four times on her hottest days.

These basic structural variances could explain perceptive differences. One study scanned the brains of men and women observing a neutral scene of a man and a woman having a conversation. The male brains' sexual areas immediately sparked—they saw it as a potential sexual rendezvous. The female brains did not have any activation in the sexual areas. The female brains saw the situation as just two people talking.

Men also have larger processors in the core of the most primitive area of the brain, which registers fear and triggers aggression—the amygdala. This is why some men can go from zero to a fistfight in a matter of seconds, while many women will try anything to defuse conflict. But the psychological stress of conflict registers more deeply in areas of the female brain. Though we live in the modern urban world, we inhabit bodies built to live in the wild, and each female brain still carries within it the ancient circuitry of her strongest foremothers, engineered for genetic success but retaining the deeply wired instincts developed in response to stress experienced in the ancient wild. Our stress responses were designed to react to physical danger and life-threatening situations. Now couple that stress response with the modern challenges of juggling the demands of home, kids, and work without enough support, and we have a situation in which women can perceive a few unpaid bills as a stress that appears to be life-threatening. This response impels the female brain to react as though the family were endangered by impending catastrophe. The male brain will not have the same perception unless the threat is of immediate, physical danger. These basic, structural variances in their brains lay the groundwork for many everyday differences in the behavior and life experiences of men and women.

Biological instincts are the keys to understanding how we are wired, and they are the keys to our success today. If you're aware of the fact that a biological brain state is guiding your impulses, you can choose not to act or to act differently than you might feel compelled. But first we have to learn to recognize how the female brain is genetically structured and shaped by evolution, biology, and culture. Without that recognition, biology becomes destiny and we will be helpless in the face of it.

Biology does represent the foundation of our personalities and behavioral tendencies. But if in the name of free will—and political correctness—we try to deny the influence of biology on the brain, we begin fighting our own nature. If we acknowledge that our biology is influenced by other factors, including our sex hormones and their flux, we can prevent it from creating a fixed reality by which we are ruled. The brain is nothing if not a talented learning machine. Nothing is completely fixed. Biology powerfully affects but does not lock in our reality. We can alter that reality and use our intelligence and determination both to celebrate and, when necessary, to change the effects of sex hormones on brain structure, behavior, reality, creativity—and destiny.

MALES AND FEMALES have the same average level of intelligence, but the female brain's reality has often been misinterpreted to mean that it is less capable in certain areas, such as math and science. In January 2005, Lawrence Summers, then president of Harvard University, shocked and enraged his colleagues—and the public—when in a speech to the National Bureau of Economic Research he said: "It does appear that on many, many different human attributes—mathematical ability, scientific ability—there is relatively clear evidence that whatever the difference in means—which can be debated—there is a difference in the standard deviation, and variability of a male and a female population. And that is true with respect to attributes that are and are not plausibly, culturally determined." The public surmised that he was saying that women are therefore innately less suited than men to be top-level mathematicians and scientists.

Judging from current research, Summers was and wasn't right. We now know that when girls and boys first hit their teen years, the difference in their mathematical and scientific capacity is nonexistent. That's where he was wrong. But as estrogen floods the female brain, females start to focus intensely on their emotions and on communication—talking on the phone and connecting with their girlfriends at the mall. At the same time, as testosterone takes over the male brain, boys grow less communicative and become obsessed about scoring—in games, and in the backseat of a car. At the point when boys and girls begin deciding the trajectories of their careers, girls start to lose interest in pursuits that require more solitary work and fewer interactions with others, while boys can easily retreat alone to their rooms for hours of computer time.

From an early age, my patient Gina had an extraordinary aptitude for math. She became an engineer but at twenty-eight years old was struggling with her desire to be in a more people-oriented career and one that would allow her to have a family life, too. She relished the mental puzzles involved in solving engineering problems, but she missed daily contact with people, so she was considering a career change. This is not an unusual conflict for women. My friend the scientist Cori Bargmann told me that many of her smartest girlfriends dropped science to go into fields that they felt were more social. These are value decisions that are actually shaped by hormonal effects on the female brain compelling connection and communication. The fact that fewer women end up in science has nothing to do with female brain deficiencies in math and science. That's where Summers really went wrong. He was right that there's a dearth of women in top-level science and engineering positions but dead wrong in implying that women do not end up in these careers because of lack of aptitude.

The female brain has tremendous unique aptitudes— outstanding verbal agility, the ability to connect deeply in friendship, a nearly psychic capacity to read faces and tone of voice for emotions and states of mind, and the ability to defuse conflict. All of this is hardwired into the brains of women. These are the talents women are born with that many men, frankly, are not. Men are born with other talents, shaped by their own hormonal reality. But that's the subject of another book.

FOR TWENTY YEARS, I've eagerly awaited progress in knowledge of the female brain and behavior as I have been treating my women patients. It was only at the turn of the millennium that exciting research started to emerge revealing how the structure, function, and chemistry of a woman's brain affect her mood, thought processes, energy, sexual drives, behavior, and well-being. This book is a user's guide to new research about the female brain and the neurobehavioral systems that make us women. It draws on my twenty years of clinical experience as a neuropsychiatrist. It culls from spectacular advances in our understanding of genetics, molecular neuroscience, fetal and pediatric endocrinology, and neurohormonal development. It presents samplings from neuropsychology, cognitive neuroscience, child development, brain imaging, and psychoneuroendocrinology. It explores primatology, animal studies, and infant observation, seeking insights into how particular behaviors are programmed into the female brain by a combination of nature and nurture.

Because of this progress, we are entering an era, finally, when women can begin to understand their distinct biology and how it affects their lives. My personal mission has been to educate interested physicians, psychologists, teachers, nurses, pharmacists, and their trainees to benefit the women and teen girls they serve. I have taken every opportunity to educate women and girls directly about their unique brain-body-behavior system and help them to be their best at every age. It is my hope that this book will benefit many more women and girls than I can personally reach in the clinic. It is my hope that the female brain will be seen and understood as the finely tuned and talented instrument that it actually is.

ONE

The Birth of the Female Brain

LEILA WAS A busy little bee, flitting around the playground, connecting with the other children whether or not she knew them. On the verge of speaking in two- and three-word phrases, she mostly used her contagious smile and emphatic nods of her head to communicate, and communicate she did. So did the other little girls. "Dolly," said one. "Shopping," said another. There was a pint-size community forming, abuzz with chatter, games, and imaginary families.

Leila was always happy to see her cousin Joseph when he joined her on the playground, but her joy never lasted long. Joseph grabbed the blocks she and her friends were using to make a house. He wanted to build a rocket, and build it by himself. His pals would wreck anything that Leila and her friends had created. The boys pushed the girls around, refused to take turns, and would ignore a girl's request to stop or give the toy back. By the end of the morning, Leila had retreated to the other end of the play area with the girls. They wanted to play house quietly together.

Common sense tells us that boys and girls behave differently. We see it every day at home, on the playground, and in classrooms. But what the culture hasn't told us is that the brain dictates these divergent behaviors. The impulses of children are so innate that they kick in even if we adults try to nudge them in another direction. One of my patients gave her three-and-a-half-year-old daughter many unisex toys, including a bright red fire truck instead of a doll. She walked into her daughter's room one afternoon to find her cuddling the truck in a baby blanket, rocking it back and forth saying, "Don't worry, little truckie, everything will be all right."

This isn't socialization. This little girl didn't cuddle her "truckie" because her environment molded her unisex brain. There is no unisex brain. She was born with a female brain, which came complete with its own impulses. Girls arrive already wired as girls, and boys arrive already wired as boys. Their brains are different by the time they're born, and their brains are what drive their impulses, values, and their very reality.

The brain shapes the way we see, hear, smell, and taste. Nerves run from our sense organs directly to the brain, and the brain does all the interpreting. A good conk on the head in the right place can mean that you won't be able to smell or taste. But the brain does more than that. It profoundly affects how we conceptualize the world—whether we think a person is good or bad, if we like the weather today or it makes us unhappy, or whether we're inclined to take care of the day's business. You don't have to be a neuroscientist to know this. If you're feeling a little down and have a nice glass of wine or a lovely piece of chocolate, your attitude can shift. A gray, cloudy day can turn bright, or irritation with a loved one can evaporate because of the way the chemicals in those substances affect the brain. Your immediate reality can change in an instant.

If chemicals acting on the brain can create different realities, what happens when two brains have different structures? There's no question that their realities will be different. Brain damage, strokes, pre-frontal lobotomies, and head injuries can change what's important to a person. They can even change one's personality from aggressive to meek or from kind to grumpy.

But it's not as if we all start out with the same brain structure. Males' and females' brains are different by nature. Think about this. What if the communication center is bigger in one brain than in the other? What if the emotional memory center is bigger in one than in the other? What if one brain develops a greater ability to read cues in people than does the other? In this case, you would have a person whose reality dictated that communication, connection, emotional sensitivity, and responsiveness were the primary values. This person would prize these qualities above all others and be baffled by a person with a brain that didn't grasp the importance of these qualities. In essence, you would have someone with a female brain.

We, meaning doctors and scientists, used to think that gender was culturally created for humans but not for animals. When I was in medical school in the 1970s and '80s, it had already been discovered that male and female animal brains started developing differently in utero, suggesting that impulses such as mating and bearing and rearing young are hardwired into the animal brain. But we were taught that for humans sex differences mostly came from how one's parents raised one as a boy or a girl. Now we know that's not completely true, and if we go back to where it all started, the picture becomes abundantly clear.

Imagine for a moment that you are in a microcapsule speeding up the vaginal canal, hitting warp drive through the cervix ahead of the tsunami of sperm. Once inside the uterus, you'll see a giant, undulating egg waiting for that lucky tadpole with enough moxie to penetrate the surface. Let's say the sperm that led the charge carries an X and not a Y chromosome. Voilà, the fertilized egg is a girl.

In the span of just thirty-eight weeks, we would see this girl grow from a group of cells that could fit on the head of a pin to an infant who weighs an average of seven and a half pounds and possesses the machinery she needs to live outside her mother's body. But the majority of the brain development that determines her sex-specific circuits happens during the first eighteen weeks of pregnancy.

Until eight weeks old, every fetal brain looks female—female is nature's default gender setting. If you were to watch a female and a male brain developing via time-lapse photography, you would see their circuit diagrams being laid down according to the blueprint drafted by both genes and sex hormones. A huge testosterone surge beginning in the eighth week will turn this unisex brain male by killing off some cells in the communication centers and growing more cells in the sex and aggression centers. If the testosterone surge doesn't happen, the female brain continues to grow unperturbed. The fetal girl's brain cells sprout more connections in the communication centers and areas that process emotion. How does this fetal fork in the road affect us? For one thing, because of her larger communication center, this girl will grow up to be more talkative than her brother. In most social contexts, she will use many more forms of communication than he will. For another, it defines our innate biological destiny, coloring the lens through which each of us views and engages the world.

READING EMOTION EQUALS READING REALITY

Just about the first thing the female brain compels a baby to do is study faces. Cara, a former student of mine, brought her baby Leila in to see us for regular visits. We loved watching how Leila changed as she grew up, and we saw her pretty much from birth through kindergarten. At a few weeks old, Leila was studying every face that appeared in front of her. My staff and I made plenty of eye contact, and soon she was smiling back at us. We mirrored each other's faces and sounds, and it was fun bonding with her. I wanted to take her home with me, particularly because I hadn't had the same experience with my son.

I loved that this baby girl wanted to look at me, and I wished my son had been so interested in my face. He was just the opposite. He wanted to look at everything else—mobiles, lights, and doorknobs— but not me. Making eye contact was at the bottom of his list of interesting things to do. I was taught in medical school that all babies are born with the need for mutual gazing because it is the key to developing the mother-infant bond, and for months I thought something was terribly wrong with my son. They didn't know back then about the many sex-specific differences in the brain. All babies were thought to be hardwired to gaze at faces, but it turns out that theories of the earliest stages of child development were female-biased. Girls, not boys, come out wired for mutual gazing. Girls do not experience the testosterone surge in utero that shrinks the centers for communication, observation, and processing of emotion, so their potential to develop skills in these areas are better at birth than boys'. Over the first three months of life, a baby girl's skills in eye contact and mutual facial gazing will increase by over 400 percent, whereas facial gazing skills in a boy during this time will not increase at all.

Baby girls are born interested in emotional expression. They take meaning about themselves from a look, a touch, every re-action from the people they come into contact with. From these cues they discover whether they are worthy, lovable, or annoy-ing. But take away the signposts that an expressive face provides and you've taken away the female brain's main touchstone for reality. Watch a little girl as she approaches a mime. She'll try with everything she has to elicit an expression. Little girls do not tolerate flat faces. They interpret an emotionless face that's turned toward them as a signal they are not doing something right. Like dogs chasing Frisbees, little girls will go after the face until they get a response. The girls will think that if they do it just right, they'll get the reaction they expect. It's the same kind of instinct that keeps a grown woman going after a narcissistic or otherwise emotionally unavailable man—"if I just do it right, he'll love me." You can imagine, then, the negative impact on a little girl's developing sense of self of the un-responsive, flat face of a depressed mother—or even one that's had too many Botox injections. The lack of facial expression is very confusing to a girl, and she may come to believe, because she can't get the expected reaction to a plea for attention or a gesture of affection, that her mother doesn't really like her. She will eventually turn her efforts to faces that are more responsive.

Anyone who has raised boys and girls or watched them grow up can see that they develop differently, especially that baby girls will connect emotionally in ways that baby boys don't. But psychoanalytic theory misrepresented this sex difference and made the assumption that greater facial gazing and the impulse to connect meant that girls were more "needy" of symbiosis with their mothers. The greater facial gazing doesn't indicate a need; it indicates an innate skill in observation. It's a skill that comes with a brain that is more mature at birth than a boy's brain and develops faster, by one to two years.

HEARING, APPROVAL AND BEING HEARD

Girls' well-developed brain circuits for gathering meaning from faces and tone of voice also push them to comprehend the social approval of others very early. Cara was surprised that she was able to take Leila out into public. "It's amazing. We can sit at a restaurant, and Leila knows, at eighteen months, that if I raise my hand she should stop reaching for my glass of wine. And I noticed that if her dad and I are arguing, she'll eat with her fingers until one of us looks over at her. Then she'll go back to struggling with a fork."

These brief interactions show Leila picking up cues from her parents' faces that her cousin Joseph likely wouldn't have looked for. A University of Texas study of twelve-month-old girls and boys showed the difference in desire and ability to observe. In this case, the child and mother were brought into a room, left alone together, and instructed not to touch an object. The mother stood off to the side. Every move, glance, and utterance was videotaped. Very few of the girls touched the forbidden object, even though their mothers never explicitly told them not to. The girls looked back at their mothers' faces ten to twenty times more than did the boys, checking for signs of approval or disapproval. The boys, by contrast, moved around the room and rarely glanced at their mothers' faces. They frequently touched the forbidden object, even though their mothers shouted, "No!" The one-year-old boys, driven by their testosterone-formed male brains, are compelled to investigate their environment, even those elements of it they are forbidden to touch.

Because their brains did not undergo a testosterone marination in utero and their communication and emotion centers were left intact, girls also arrive in the world better at reading faces and hearing human vocal tones. Just as bats can hear sounds that even cats and dogs cannot, girls can hear a broader range of sound frequency and tones in the human voice than can boys. Even as an infant, all a girl needs to hear is a slight tightening in her mother's voice to know she should not be opening the drawer with the fancy wrapping paper in it. But you will have to restrain the boy physically to keep him from destroying next Christmas's packages. It's not that he's ignoring his mother. He physically cannot hear the same tone of warning.

A girl is also astute at reading from facial expression whether or not she's being listened to. At eighteen months, Leila could not be kept quiet. We couldn't understand anything she was trying to tell us, but she waddled up to each person in the office and unloosed a stream of words that seemed very important to her. She tested for agreement in each of us. If we appeared even the tiniest bit disinterested, or broke eye contact for a second, she put her hands on her hips, stomped her foot, and grunted in indignation. "Listen!" she yelled. No eye contact meant to her that we were not listening. Cara and her husband, Charles, were worried that Leila seemed to insist on being included in any conversation at home. She was so demanding that they thought they had spoiled her. But they hadn't. It was just their daughter's brain searching for a way to validate her sense of self.

Whether or not she is being listened to will tell a young girl if others take her seriously, which in turn goes to the growth of her sense of a successful self. Even though her language skills aren't developed, she understands more than she expresses, and she knows—before you do—if your mind has wandered for an instant. She can tell if the adult understands her. If the adult gets on the same wavelength, it actually creates her sense of self as being successful or important. If she doesn't connect, her sense is of an unsuccessful self. Charles in particular was surprised by how much focus it took to keep up the relationship with his daughter. But he saw that, when he listened attentively, she began to develop more confidence.

EMPATHY

This superior brain wiring for communication and emotional tones plays out early in a baby girl's behavior. Years later Cara couldn't understand why her son didn't settle down as quickly when she picked him up as her daughter, Leila, had. She thought it was just temperament, a fussier personality. But likely it was also the sex difference in hardwiring in the brain for empathy. The baby girl is able to resonate more easily with her mother and respond quickly to soothing behavior, stopping her fussing and crying. Observations made during a study at Harvard Medical School found that baby girls do this better with their mothers than do boys.

Another study showed that typical female newborns less than twentyfour hours old respond more to the distressed cries of another baby—and to the human face—than male newborns do. Girls as young as a year old are more responsive to the distress of other people, especially those who look sad or hurt. I was feeling a little down one day and mentioned it to Cara. Leila, at eighteen months, picked up on my tone of voice. She climbed onto my lap and played with my earrings, hair, and glasses. She held my face in her hands, looked right into my eyes, and I felt better immediately. That little girl knew exactly what she was doing.

At this stage Leila was in the hormone phase of what is called infantile puberty, a period that lasts only nine months for boys, but is twenty-four months long for girls. During this time, the ovaries begin producing huge amounts of estrogen—comparable to the level of an adult female—that marinate the little girl's brain. Scientists believe these infantile estrogen surges are needed to prompt the development of the ovaries and brain for reproductive purposes. But this high quantity of estrogen also stimulates the brain circuits that are rapidly being built. It spurs the growth and development of neurons, further enhancing the female brain circuits and centers for observation, communication, gut feelings, even tending and caring. Estrogen is priming these innate female brain circuits so that this little girl can master her skills in social nuance and promote her fertility. That's why she was able to be so emotionally adept while still in diapers.

INHERITING MORE THAN MOM'S GENES

Because of her ability to observe and feel emotional cues, a girl actually incorporates her mother's nervous system into her own. Sheila came to me wanting some help dealing with her kids. With her first husband she had two daughters, Lisa and Jennifer. When Lisa was born, Sheila was still happy and content in her first marriage. She was an able and highly nurturing mother. By the time Jennifer was born, eighteen months later, circumstances had changed considerably. Her husband had become a flagrant philanderer. Sheila was being harassed by the husband of the woman he was having an affair with. And things got worse. Sheila's unfaithful husband had a powerful and rich father, who threatened to have the children kidnapped if she tried to leave the state to be with her own family for support.

It was in this stressful environment that Jennifer spent her infancy. Jennifer became suspicious of everyone and by age six started telling her older sister that their kind and beloved new stepfather was certainly cheating on their mother. Jennifer was sure of it and repeated her suspicions frequently. Lisa, finally went to their mom and asked if it were true. Their new stepfather was one of those men who just didn't have it in him to cheat, and Sheila knew it. She couldn't figure out why her younger daughter had become so anxiously fixated on the imagined infidelity of her new husband. But Jennifer's nervous system had imprinted the unsafe perceptual reality of her earliest years, so even good people seemed unreliable and threatening. The two sisters were raised by the same mother but under different circumstances, so one daughter's brain circuits had incorporated a nurturing, safe mom and the other's a fearful, anxious one.

The "nervous system environment" a girl absorbs during her first two years becomes a view of reality that will affect her for the rest of her life. Studies in mammals now show that this early stress versus calm incorporation—called epigenetic imprinting—can be passed down through several generations. Research in mammals by Michael Meaney's group has shown that female offspring are highly affected by how calm and nurturing their mothers are. This relation has also been shown in human females and nonhuman primates. Stressed mothers naturally become less nurturing, and their baby girls incorporate stressed nervous systems that change the girls' perception of reality. This isn't about what's learned cognitively—it's about what is absorbed by the cellular microcircuitry at the neurological level. This may explain why some sisters can have amazingly different outlooks. It appears that boys may not incorporate so much of their *mothers'* nervous system.

Neurological incorporation begins during pregnancy. Maternal stress during pregnancy has effects on the emotional and stress hormone reactions, particularly in female offspring. These effects were measured in goat kids. The stressed female kids ended up startling more easily and being less calm and more anxious than the male kids after birth. Furthermore, female kids who were stressed in utero showed a great deal more emotional distress than female kids who weren't. So if you're a girl about to enter the womb, plan to be born to an unstressed mom who has a calm, loving partner and family to support her. And if you are a mom-to-be carrying a female fetus, take it easy so that your daughter will be able to relax.

DON'T FIGHT

So why is a girl born with such a highly tuned machine for reading faces, hearing emotional tones in voices, and responding to unspoken cues in others? Think about it. A machine like that is built for connection. That's the main job of the girl brain, and that's what it drives a female to do from birth. This is the result of millennia of genetic and evolutionary hardwiring that once had—and probably still has—real consequences for survival. If you can read faces and voices, you can tell what an infant needs. You can predict what a bigger, more aggressive male is going to do. And since you're smaller, you probably need to band with other females to fend off attacks from a ticked off caveman—or cavemen.

If you're a girl, you've been programmed to make sure you keep social harmony. This is a matter of life and death to the brain, even if it's not so important in the twenty-first century. We could see this in the behavior of three-and-a-half-year-old twin girls. Every morning the sisters climbed on each other's dressers to get to the clothes hanging in their closets. One girl had a pink two-piece outfit, and the other had a green two-piece outfit. Their mother giggled every time she'd see them switch the tops—pink pants with a green top and green pants with a pink top. The twins did it without a fight. "Can I borrow your pink top? I'll give it back later, and you can have my green top" was how the dialogue went. This would not be a likely scenario if one of the twins were a boy. A brother would have grabbed the shirt he wanted, and the sister would have tried to reason with him, though she would have ended up in tears because his language skills simply wouldn't have been as advanced as hers.

Typical non-testosteronized, estrogen-ruled girls are very invested in preserving harmonious relationships. From their earliest days, they live most comfortably and happily in the realm of peaceful interpersonal connections. They prefer to avoid conflict because discord puts them at odds with their urge to stay connected, to gain approval and nurture. The twenty-four-month estrogen bath of girls' infantile puberty reinforces the impulse to make social bonds based on communication and compromise. It happened with Leila and her new friends on the playground. Within a few minutes of meeting they were suggesting games, working together, and creating a little community. They found a common ground that led to shared play and possible friendship. And remember Joseph's noisy entrance? That usually wrecked the day and the harmony sought out by the girls' brains.

It is the brain that sets up the speech differences—the genderlects—of small children, which Deborah Tannen has pointed out. She noted that in studies of the speech of two-to five-year-olds, girls usually make collaborative proposals by starting their sentences with "let's"—as in "Let's play house." Girls, in fact, typically use language to get consensus, influencing others without telling them directly what to do. When Leila hit the playground, she said "Shopping" as a suggestion for how she and her companions might play together. She looked around and waited for a response instead of forging ahead. The same thing happened when another little girl said "Dolly." As has been observed in studies, girls participate jointly in decision making, with minimal stress, conflict, or displays of status. They often express agreement with a partner's suggestions. And when they have ideas of their own, they'll put them in the form of questions, such as "I'll be the teacher, okay?" Their genes and hormones have created a reality in their brains that tells them social connection is at the core of their being.

Boys know how to employ this affiliative speech style, too, but research shows they typically don't use it. Instead, they'll generally use language to command others, get things done, brag, threaten, ignore a partner's suggestion, and override each other's attempts to speak. It was never long after Joseph's arrival on the playground that Leila ended up in tears. At this age boys won't hesitate to take action or grab something they desire. Joseph took Leila's toys whenever he wanted and usually destroyed whatever Leila and the other girls were making. Boys will do this to one another—they are not concerned about the risk of conflict. Competition is part of their makeup. And they routinely ignore comments or commands given by girls.

The testosterone-formed boy brain simply doesn't look for social connection in the same way a girl brain does. In fact, disorders that inhibit people from picking up on social nuance—called autism spectrum disorders and Asperger's syndrome—are eight times more common in boys. Scientists now believe that the typical male brain, with only one dose of X chromosome (there are two X's in a girl), gets flooded with testosterone during development and somehow becomes more easily socially handicapped. Extra testosterone in people with these disorders may be killing off some of the brain's circuits for emotional and social sensitivity.

SHE WANTS COMMUNITY, BUT ONLY ON HER TERMS

By age two and a half, infantile puberty ends and a girl enters the calmer pastures of the juvenile pause. The estrogen stream coming from the ovaries has been temporarily stopped; how, we don't yet know. But we do know that the levels of estrogen and testosterone become very low during the childhood years in both boys and girls—although girls still have six to eight times more estrogen than boys. When women talk about "the girl they left behind," this is the stage they are usually referring to. This is the quiet period before the full-volume rock 'n' roll of puberty. It's the time when a girl is devoted to her best friend, when she doesn't usually enjoy playing with boys. Research shows that this is true for girls between the ages of two and six in every culture that's been studied.

I met my first playmate, Mikey, when I was two and a half and he was almost three. My family had moved into a house next door to Mikey's on Quincy Street in Kansas City, and our back-yards adjoined each other. The sandbox was in our yard, and the swing set straddled the invisible line that divided our properties.

Our mothers, who soon became friends, saw the advantage of their two kids playing with each other while they chatted or took turns watching us. According to my mother, almost every time Mikey and I played in the sandbox, she would have to rescue me because he would inevitably grab my toy shovel or pail while refusing to let me touch his. I would wail in protest, and Mikey would scream and hurl sand at us as his mother tried to pry my toys away from him.

Both our moms tried again and again, because they liked spending time together. But nothing Mikey's mother did—scolding him, reasoning with him about the merits of sharing, taking away privileges, imposing various punishments—could persuade him to change his behavior. My mother eventually had to look beyond our block to find me other playmates, girls who sometimes grabbed but always could be reasoned with, who might use words to be hurtful but never raised a hand to hit or punch. I had begun to dread the daily battles with Mikey, and I was happy about the change.

The cause for this preference for same-sex playmates remains largely unknown, but scientists speculate that basic brain differences may be one reason. Girls' social, verbal, and relationship skills develop years earlier than boys'. That their styles of communication and interaction are completely different is probably a result of these brain variations. Typical boys enjoy wrestling, mock fighting, and rough play with cars, trucks, swords, guns, and noisy—preferably explosive—toys. They also tend to threaten others and get into more conflict than girls beginning as early as age two, and they're less likely to share toys and take turns than are female children. Typical girls, by contrast, don't like rough play—if they get into too many tussles, they'll just stop playing. According to Eleanor Maccoby, when girls get pushed around too much by boys their age—who are just having fun—they will retreat from the space and find another game to play, preferably one that doesn't involve any high-spirited boys.

Studies show girls take turns twenty times more often than boys, and their pretend play is usually about interactions in nurturing or caregiving relationships. Typical female brain development underlies this behavior. Girls' social agenda, expressed in play and determined by their brain development, is to form close, one-on-one relationships. Boys' play, by contrast, is usually not about relationships—it's about the game or toy itself as well as social rank, power, defense of territory, and physical strength.

In a 2005 study done in England, little boys and girls were compared at four years of age on the quality of their social relationships. This comparison included a popularity scale on which they were judged by how many other children wanted to play with them. Little girls won hands down. These same four-year-old children had had their testosterone levels measured in utero between ages twelve and eighteen weeks, while their brains were developing into a male or a female design. Those with the lowest testosterone exposure had the highest quality social relationships at four years old. They were the girls.

Studies of nonhuman female primates also provide clues that these sex differences are innate and require the right hormone-priming actions. When researchers block estrogen in young female primates during infantile puberty, the females don't develop their usual interest in infants. Moreover, when researchers inject female primate fetuses with testosterone, the injected females end up liking more rough-and-tumble play than do average females. This is also true in humans. Though we have not performed experiments to block estrogen in little girls, or injected testosterone into human fetuses, we can see this brain effect of testosterone at work in the rare enzyme deficiency called congenital adrenal hyperplasia (CAH), which occurs in about one out of every ten thousand infants.

Emma did not want to play with dolls. She liked trucks and jungle gyms and sets to build things with. If you asked her at two and a half years old if she was a boy or a girl, she'd tell you she was a boy and she'd punch you. She'd get a running start, and "the little linebacker," as her mother called her, would knock over anyone who came into the room. She played catch with stuffed animals, though she threw them so hard it was tough to hang on to them. She was rough, and the girls at preschool didn't want to play with her. She was also a little behind the other girls in language development. Yet Emma liked dresses and loved when her aunt styled her hair. Her mother, Lynn, an avid cyclist, athlete, and science teacher, wondered, when she brought Emma in to see me, if her being a jock had influenced her daughter's behavior. Most of the time, a girl like Emma would be the one in ten who is simply a tomboy. In this case, Emma had CAH.

Congenital adrenal hyperplasia causes fetuses to produce large amounts of testosterone, the sex and aggression hormone, from their adrenal glands starting at about eight weeks after conception—the very moment their brains begin to take shape into the male or female design. If we look at genetic females whose brains are exposed to surges of testosterone during this period, we see that these girls' behavior and presumably brain structures are more similar to those of males than to those of females. I say "presumably" because a toddler's brain isn't so easy to study. Can you imagine a two-year-old sitting still for a couple of hours in an MRI scanner without being sedated? But we can deduce a lot from behavior.

The study of congenital adrenal hyperplasia provides evidence that testosterone erodes the normally robust brain structures in girls. At one year old, CAH girls make measurably less eye contact than other girls the same age. As these testosterone-exposed girls get older, they are far more inclined to scuffling, roughhousing, and fantasy play about monsters or action heroes than to pretending to take care of their dolls or dressing up in princess costumes. They also do better than other girls on spatial tests, scoring similarly to boys, while they do less well on tests that tap verbal behavior, empathy, nurturing, and intimacy—traits that are typically female. The implications are that the male and female brains' wiring for social connection is significantly affected not just by genes but by the amount of testosterone that gets into the fetal brain. Lynn was relieved to have a scientific reason for some of her daughter's behaviors, since no one had taken the time to explain to her what happens in the CAH brain.

GENDER EDUCATION

Nature certainly has the strongest hand in launching sex-specific behaviors, but experience, practice, and interaction with others can modify neurons and brain wiring. If you want to learn to play the piano, you must practice. Every time you practice, your brain assigns more neurons to that activity, until finally you have laid new circuits between these neurons so that, when you sit down at the bench, playing is second nature.

As parents, we naturally respond to our children's preferences. We will repeat, sometimes ad nauseam, the activity—Mommy's smile or the noisy whistle of a wooden train—that makes our little one giggle or grin. This repetition strengthens those neurons and circuits in the baby's brain that process and respond to whatever initially captivated her or his attention. The cycle continues, and children thus learn the customs of their gender. Since a little girl responds so well to faces, chances are Mom and Dad will make a lot of faces and she'll get even better at responding. She'll be engaged in an activity that reinforces her face-studying skill, and her brain will assign more and more neurons to that activity. Gender education and biology collaborate to make us who we are.

Adult expectations for girls' and boys' behavior play an important role in shaping brain circuits, and Wendy could have blown it for her daughter Samantha if she had given in to her own preconceptions about girls being more fragile and less adventurous than boys. Wendy told me that the first time Samantha climbed the jungle gym ladder to go down the slide by herself, she immediately looked back at Wendy for permission. If she had sensed disapproval or fear in her mother's facial expression, she probably would have stopped, climbed back down, and asked for her mother's help—as would 90 percent of little girls. When Wendy's son was that age, he would never have bothered checking for her reaction, not caring if Wendy disapproved of this step of independence. Samantha obviously felt ready to take this "big girl" leap, so Wendy managed to squelch her fear and give her daughter the approval she needed. She says she wishes she had had a camera to record the moment Samantha landed with a bump at the bottom. Her face lit up with a grin that expressed her pride and excitement, and she immediately ran over to her mother and gave her a big hug.

The brain's first organizing principle is clearly genes plus hormones, but we can't ignore the further sculpting of the brain that results from our interactions with other people and with our environment. A parent's or caregiver's tone of voice, touch, and words help organize an infant's brain and influence a child's version of reality.

Scientists still don't know exactly how much reshaping can occur to the brain nature gave us. It runs against the grain of intuition, but some studies show that male and female brains may have different genetic susceptibility to environmental influences. Either way, we know enough to see that the fundamentally misconceived nature versus nurture debate should be abandoned: child development is inextricably both.

THE BOSSY BRAIN

If you're the parent of a little girl, you know firsthand that she isn't always as obedient and good as the culture would have us believe she should be. Many parents have had their expectations dashed when it came to their daughter getting what she wanted.

"Okay, Daddy, now the dollies are going to lunch, so we have to change their clothes," Leila said to her father, Charles, who dutifully changed the outfits—into party clothes. "Daddy! No," Leila screamed. "Not the party dress! The lunch outfits! And they don't talk like that. You're supposed to say what I told you to say. Now say it right."

"All right, Leila. I'll do it. But tell me, why do you like to play dolls with me instead of with Mommy?"

"Because, Daddy, you play the way I tell you to." Charles was a little thrown by this response. And he and Cara were taken aback by Leila's chutzpah.

Not all is perfectly calm during the juvenile pause. Little girls don't usually exhibit aggression via rough-and-tumble play, wrestling, and punching the way little boys do. Girls may have, on average, better social skills, empathy, and emotional intelligence than boys—but don't be fooled. This doesn't mean that girls' brains aren't wired to use everything in their power to get what they want, and they can turn into little tyrants to accomplish their goals. What are those goals as dictated by the little girl's brain? To forge connection, to create community, and to organize and orchestrate a girl's world so that she's at the center of it. This is where the female brain's aggression plays out—it protects what's important to it, which is always, inevitably, relationship. But aggression can push others away, and that would undermine the goal of the female brain. So a girl walks a fine line between making sure she's at the center of her world of relationships and risking pushing those relationships away.

Remember the wardrobe sharing twins? When one asked the other to borrow the pink shirt in trade for the green, she set it up so that if the other sister said no she'd be considered mean. Instead of grabbing the shirt, she used her best skill set—language—to get what she wanted. She was counting on her sister's not wanting to be seen as selfish, and indeed her sister gave up the pink shirt. She got what she wanted without sacrificing the relationship. This is aggression in pink. Aggression means survival for both sexes, and both sexes have brain circuits for it. It's just more subtle in girls, perhaps reflecting their unique brain circuitry.

The social and scientific view of innate good behavior in girls is a misguided stereotype born out of the contrast with boys. In comparison, girls come out smelling like roses. Women don't need to lay one another out, so of course they seem less aggressive than males. By all standards, men are on average twenty times more aggressive than women, something that a quick look around the prison system will confirm. I almost left aggression out of this book, after being lulled into a warm glow of communicative and social female brain circuits. I was nearly fooled by the female aversion to conflict into thinking that aggression simply wasn't part of our makeup.

Cara and Charles didn't know what to do about Leila's bossiness. It didn't end with telling her father how to play dolls. She screamed when her friend Susie painted a yellow clown instead of a blue one as she had ordered, and heaven forbid if a conversation at the dinner table didn't include Leila. Her female brain was demanding that she be part of whatever communication or connection was taking place in her presence. Being left out was more than her girl circuits could bear. To her Stone Age brain— and face it, we're all still cave people inside— being left out could mean death. I explained this to Cara and Charles, and they decided to wait out this phase instead of trying to change Leila's behavior—within reason, of course.

I didn't want to tell Cara and Charles that what Leila was putting them through was nothing. Her hormones were steady, they were at a low point, and her reality was fairly stable. When the hormones turn back on and the juvenile pause comes to an end, Cara and Charles won't have just Leila's bossy brain to deal with. Her risk-taking brain will have the stops pulled out. It will drive her to ignore her parents, entice a mate, leave home, and make something different out of herself. Teen girl reality will explode, and every trait established in the female brain during girlhood—communication, social connection, desire for approval, reading faces for cues as to what to think or feel—will intensify. This is the time when a girl becomes most communicative with her girlfriends and forms tightly knit social groups in order to feel safe and protected. But with this new estrogen-driven reality, aggression also plays a big role. The teen girl brain will make her feel powerful, always right, and blind to consequences. Without that drive, she'll never be able to grow up, but getting through it, especially for the teen girl, isn't easy. As she begins to experience her full "girl power," which includes premenstrual syndrome, sexual competition, and controlling girl groups, her brain states can often make her reality, well, a little hellish.

TWO

Teen Girl Brain

DRAMA, DRAMA, DRAMA. That's what's happening in a teen girl's life and a teen girl's brain. "Mom, I so totally can't go to school. I just found out Brian likes me and I have a huge zit and no concealer. OMG! How can you even think I'll go?" "Homework? I told you I'm not doing any more until you promise to send me away to school. I can't stand living with you for one more minute." "No, I'm not done talking to Eve. It has *not* been two hours, and I'm not getting off the phone." This is what you get if you have the modern version of the teen girl brain living in your house.

The teenage years are a turbulent time. The teen girl's brain is sprouting, reorganizing and pruning neuronal circuits that drive the way she thinks, feels, and acts—and obsesses over her looks. Her brain is unfolding ancient instructions on how to be a woman. During puberty, a girl's entire biological raison d'être is to become sexually desirable. She begins judging herself against her peers and media images of other attractive females. This brain state is created by the surge of new hormones on top of the ancient female genetic blueprint.

Attracting male attention is a newfound and exciting form of self-expression for my friend Shelly's teenage daughters, and the high-octane estrogen coursing through their brain pathways fuels their obsession. The hormones that affect their responsivity to social stress are going sky high, which is where they get their off-the-wall ideas—and clothing choices—and why they are constantly staring at themselves in the mirror. They are almost exclusively interested in their appearance, specifically whether the boys who populate their real and fantasy worlds will find them attractive. Thank goodness, says Shelley, they have three bathrooms in their home, because her girls spend hours in front of the mirror, inspecting pores, plucking eyebrows, wishing the butts they see would shrink, their breasts grow larger and waists get smaller, all to attract boys. Girls would likely be doing some version of this whether the media were there to influence their self-image or not. Hormones would be driving their brains to develop these impulses even if they didn't see skinny actresses and models on the cover of every magazine. They would be obsessing over whether or not boys thought they looked good because their hormones create the reality in their brains that being attractive to boys is the most important thing.

Their brains are hard at work rewiring themselves, and this is why conflicts will increase and become more intense as teen girls struggle for independence and identity. Who are they anyway? They are developing the parts of themselves that most make them women—their strength for communicating, forming social bonds, and nurturing those around them. If parents understand the biological changes happening in the teen girl brain circuits, they can support their daughters' self-esteem and well-being during these rocky years.

RIDING THE ESTROGEN-PROGESTERONE WAVES

The smooth sailing of girlhood is over. Now parents find themselves walking on eggshells around a moody, temperamental, and resistant child. All of this drama is because the girlhood or juvenile pause has ended, and their daughter's pituitary gland has sprung to life as the chemical brakes are taken off her pulsing hypothalamic cells, which have been held in check since toddlerhood. This cellular release sparks the hypothalamic-pituitary-ovarian system into action. It is the first time since

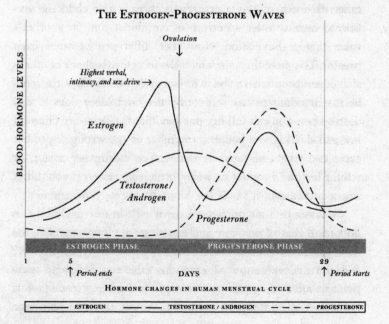

THE ESTROGEN-PROGESTERONE WAVES

HORMONE CHANGES IN HUMAN MENSTRUAL CYCLE

59

infantile puberty that their daughter's brain will be marinated in high levels of estrogen. In fact, it is the first time that her brain will experience estrogen-progesterone surges that come in repeated monthly waves from her ovaries. These surges will vary day to day and week to week.

The rising tide of estrogen and progesterone starts to fuel many circuits in the teen girl's brain that were laid down in fetal life. These new hormonal surges assure that all of her female-specific brain circuits will become even more sensitive to emotional nuance, such as approval and disapproval, acceptance and rejection. And as her body blossoms, she may not know how to interpret the newfound sexual attention—are those stares of approval or disapproval? Are her breasts the right ones or the wrong ones? On some days her self-confidence is strong, and on other days it hangs by a precarious thread. As a child she was able to hear a wider spectrum of emotional tone in another's voice than a boy could. Now that difference becomes even greater. The filter through which she feels the feedback of others also depends on where she is in her cycle—some days the feedback will reinforce her self-confidence, and other days it will destroy her. You can tell her one day that her jeans are cut a bit low and she'll ignore you. But catch her on the wrong day of her cycle and what she hears is that you're calling her a slut, or telling her she's too fat to wear those jeans. Even if you didn't say or intend this, it's how her brain interprets your comment.

We know that many parts of the female brain—including an important seat of memory and learning (the hippocampus), the main center for control of the body's organs (the hypothalamus), and the master center of emotions (the amygdala) are particularly affected by this new estrogen and progesterone fuel. It sharpens critical thinking and fine-tunes emotional responsivity.

These enhanced brain circuits will stabilize into their adult shape by late puberty and into early adulthood. At the same time, we now know that the estrogen and progesterone surges start making the adolescent female brain, especially the hippocampus, experience weekly changes in sensitivity to stress that will continue until she passes through menopause.

Researchers at the Pittsburg Psychobiologic Studies Center studied normal seven- to sixteen-year-olds as they progressed through puberty, testing their stress responsivity and their daily levels of cortisol. The girls showed more intense responses, while boys' stress responsiveness dropped. Females' bodies and brains react to stress differently than do males' once they have entered puberty. Fluctuating estrogen and progesterone in the brain is responsible for this opposite stress responsivity in the hippocampus of females. Males and females become reactive to different kinds of stress. Girls begin to react more to relation-ship stresses and boys to challenges to their authority. Relationship conflict is what drives a teen girl's stress system wild. She needs to be liked and socially connected; a teen boy needs to be respected and higher in the male pecking order.

The girl's brain circuits are arranged and fueled by estrogen to respond to stress with nurturant activities and the creation of protective social networks. She hates relationship conflict. Her brain's stress response is massively triggered by social rejection. The ebb and flow of estrogen during the menstrual cycle changes this sensitivity to psychological and social stress on a weekly basis. The first two weeks of the cycle, when estrogen is high, a girl is more likely to be socially interested and relaxed with others. In the last two weeks of the cycle, when progesterone is high and estrogen is lower, she is more likely to react with increased irritability and will want to be left alone.

Estrogen and progesterone reset the brain's stress response each month. A girl's self-confidence may be high one week but on thin ice the next.

During the juvenile pause of childhood, when estrogen levels are stable and low, a girl's stress system is calmer and more constant. Once estrogen and progesterone levels climb at puberty, her responsivity to both stress and pain start to rise, all marked by new reactions in the brain to the stress hormone cortisol. She's easily stressed, high-strung, and she starts looking for ways to chill out.

SO HOW DOES SHE CALM DOWN?

I was teaching a class of fifteen-year-olds about brain differences between males and females, and I asked the boys and girls to come up with some questions that they'd always wanted to ask each other. The boys asked, "Why do girls go to the bathroom together?" They assumed that the answer would involve something sexual, but the girls replied: "It's the only private place at school we can go to *talk*!" Needless to say, the boys couldn't ever imagine saying to another guy: "Hey, want to go to the bathroom together?"

That scene captures a pivotal brain difference between males and females. As we saw in Chapter 1, the circuits for social and verbal connection are more naturally hardwired in the typical female brain than in the typical male. It is during the teen years that the flood of estrogen in girls' brains will activate oxytocin and sex-specific female brain circuits, especially those for talking, flirting, and socializing. Those high school girls hanging out in the bathroom are cementing their most important relationships—with other girls.

Many women find biological comfort in one another's company, and language is the glue that connects one female to another. No surprise, then, that some verbal areas of the brain are larger in women than in men and that women, on average, talk and listen a lot more than men. The numbers vary, but on average girls speak two to three times more words per day than boys. We know that young girls speak earlier and by the age of twenty months have double or triple the number of words in their vocabularies than do boys. Boys eventually catch up in their vocabulary but not in speed. Girls speak faster on average, especially when they are in a social setting. Men haven't always appreciated that verbal edge. In Colonial America, women were put in the town stocks with wooden clips on their tongues or tortured by the "dunking stool," held underwater and almost drowned—punishments that were never imposed on men—for the crime of "talking too much." Even among our primate cousins, there's a big difference in the vocal communication of males and females. Female rhesus monkeys, for instance, learn to vocalize much earlier than do males and use every one of the seventeen vocal tones of their species all day long, every day, to communicate with one another. Male rhesus monkeys, by contrast, learn only three to six tones, and once they're adults, they'll go for days or even weeks without vocalizing at all. Sound familiar?

And why do girls go to the bathroom to talk? Why do they spend so much time on the phone with the door closed? They're trading secrets and gossiping to create connection and intimacy with their female peers. They're developing close-knit cliques with secret rules. In these new groups, talking, telling secrets, and gossiping, in fact, often become girls' favorite activities—their tools to navigate and ease the ups and downs and stresses of life.

I could see it in Shana's face. Her mother was complaining that she couldn't get her fifteen-year-old to concentrate on work, or even a conversation about school. Forget keeping her at the table for dinner. Shana had an almost drugged look sitting in my waiting room while she anticipated the next text message from her girlfriend Parker. Shana's grades hadn't been great, and she was becoming a bit of a behavior problem at school, so she wasn't allowed to go over to her friend's. Her mother, Lauren, had also denied her use of the cell phone and the computer, but Shana's reaction to being cut off from her friends was so over the top—she screamed, slammed doors, and started wrecking her room—that Lauren relented and gave her twenty minutes per day on the cell phone to make contact. But since she couldn't talk in private, Shana resorted to text messaging.

There is a biological reason for this behavior. Connecting through talking activates the pleasure centers in a girl's brain. Sharing secrets that have romantic and sexual implications activates those centers even more. We're not talking about a small amount of pleasure. This is huge. It's a major dopamine and oxytocin rush, which is the biggest, fattest neurological reward you can get outside of an orgasm. Dopamine is a neuro-chemical that stimulates the motivation and pleasure circuits in the brain. Estrogen at puberty increases dopamine and oxytocin production in girls. Oxytocin is a neurohormone that triggers and is triggered by intimacy. When estrogen is on the rise, a teen girl's brain is pushed to make even more oxytocin—and to get even more reinforcement for social bonding. At midcycle, during peak estrogen production, the girl's dopamine and oxytocin level is likely at its highest, too. Not only her verbal output is at its maximum but her urge for intimacy is also peaking. Intimacy releases more oxytocin, which reinforces the desire to connect, and connecting then brings a sense of pleasure and well-being.

Both oxytocin and dopamine production are stimulated by ovarian estrogen at the onset of puberty—and for the rest of a woman's fertile life. This means that teen girls start getting even more pleasure from connecting and bonding—playing with each other's hair, gossiping, and shopping together—than they did before puberty. It's the same kind of dopamine rush that coke or heroin addicts get when they do drugs. The combination of dopamine and oxytocin forms the biological basis of this drive for intimacy with its stress-reducing effect. If your teenage daughter is constantly talking on the phone or instant-messaging with her friends, it's a girl thing, and it is helping her through stressful social changes. But you don't have to let her impulses dictate your family life. It took Lauren months of negotiation to get Shana to sit through a family dinner without text-messaging the world. Because the teen girl's brain is so well-rewarded for communication, it's a tough habit for you to curb.

BOYS WILL BE BOYS

We know that girls' estrogen levels climb at puberty and flip the switches in their brains to talk more, interact with peers more, think about boys more, worry about appearance more, stress out more, and emote more. They are driven by a desire for connection with other girls—and with boys. Their dopamine and oxytocin rush from talking and connecting keeps them motivated to seek out these intimate connections. What they don't know is that this is their own special girl reality. Most boys don't share this intense desire for verbal connection, so attempts at verbal intimacy with their male contemporaries can be met with disappointing results. Girls who expect their boyfriends to chat with them the way their girlfriends do are in for a big surprise. Phone conversations can have painful lulls while she waits for him to say something. The best she can often hope for is that he is an attentive listener. She may not realize he's just bored and wants to get back to his video game.

This difference may also be at the core of the major disappointment women feel all their lives with their marriage partners—he doesn't feel like being social, he doesn't crave long talks. But it's not his fault. When he is a teen, his testosterone levels begin soaring off the charts, and he "disappears into adolescence," a phase used by one psychologist friend of mine to describe why her fifteen-year-old son never wants to talk to her anymore, takes refuge with his buddies in person or online gaming, and visibly cringes at the thought of a family dinner or outing. More than anything, he wants to be left alone in his room.

Why do previously communicative boys become so taciturn and monosyllabic that they verge on autistic when they hit their teens? The testicular surges of testosterone marinate the boys' brains. Testosterone has been shown to decrease talking as well as interest in socializing—except when it involves sports or sexual pursuit. In fact, sexual pursuit and body parts become pretty much obsessions.

When I was teaching the class of fifteen-year-olds and it was time for the girls to ask their questions of boys, they wanted to know this: "Do you prefer girls who have a little hair or a lot of hair?" I thought they meant hairstyles, as in long hair versus a shorter cut. But I quickly realized that they were referring to the boys' preference for a lot or a little pubic hair. The boys resoundingly responded "No hair at all." So let's not mince words here. Young teen boys are often totally, single-mindedly consumed with sexual fantasies, girls' body parts, and the need to masturbate. Their reluctance to talk to adults comes out of magical thinking that grown-ups will read between their spoken lines and the look in their eyes and know that the subject of sex has taken them over, mind, body, and soul.

A teenage boy feels alone in and ashamed by his thoughts. Until his buddies start to joke and comment about girls' bodies, he thinks he is the only one consumed by such intense sexual fantasies and the constant fear that someone will notice the erections over which he seems to have no control. Compelling masturbatory frenzies overwhelm him many times a day. He lives in fear of being "found out." He's even more wary of verbal intimacy with girls, though he dreams of other kinds of intimacy with them day in and day out. For a few of the teen years, the teen girl brain and the teen boy brain have seriously different priorities when it comes to being close.

FEAR OF CONFLICT

Studies indicate that girls are motivated—on a molecular and a neurological level—to ease and even prevent social conflict. Maintaining the relationship at all costs is the female brain's goal. This may be especially true in the teenage female brain.

I remember when my friend Shelley's oldest teenage daughter, Elana, had sleepovers nearly every night with her best friend, Phyllis—and on the nights they didn't, they talked on the phone until they had to go to sleep. They planned their outfits, talked about crushes on boys, and watched TV together over the phone. One day Phyllis started bad-mouthing a less popular girl in class, who had been close friends with Elana in grade school. Her meanness made Elana uncomfortable and angry, but as she thought about confronting Phyllis, her mind and body were seared by a wave of anxiety. It became real to her that if she made even a hint of criticism to Phyllis's face, a fight that would spell the end of the friendship could result. Instead of risking the loss of her friendship with Phyllis, Elana decided to say nothing.

This is a tape that plays in the brain of every woman at the thought of conflict, even a small disagreement. The female brain has a far more negative alert reaction to relationship conflict and rejection than does the male brain. Men often enjoy interpersonal conflict and competition; they even get a positive boost from it. In women, conflict is more likely to set in motion a cascade of negative chemical reactions, creating feelings of stress, upset, and fear. Just the thought that there might be a conflict will be read by the female brain as threatening the relationship, and bring with it the real concern that the next conversation she has with her friend will be their last.

When a relationship is threatened or lost, the bottom drops out of the level of some of the female brain's neurochemicals—such as serotonin, dopamine, and oxytocin (the bonding hormone)—and the stress hormone cortisol takes over. A woman starts feeling anxious, bereft, and fearful of being rejected and left alone. Soon she begins to jones for that good intimacy drug, oxytocin. She gets a feeling of closeness from the flood of oxytocin, which is boosted by social contact. But the minute that social contact is gone and the oxytocin and dopamine bottom out, she's in emotional trouble.

As soon as a woman gets her feelings hurt, the hormonal shift sets off a fearful fantasy that the relationship will be over. This is why Elana decided to let Phyllis's mean comment about her old friend go so she didn't have to risk the fight that might end the friendship. That's the fearful reality playing out in the female brain. This is why the breakup of a friendship, or just the thought of social isolation, is so stressful, especially among girl teens. Many brain circuits are geared to monitor closeness, and when closeness is threatened, the brain sounds the abandonment alarm loudly. Robert Josephs at the University of Texas has concluded that men's self-esteem derives more from their ability to maintain independence from others, while women's self-esteem is maintained, in part, by the ability to sustain intimate relationships with others. As a result, perhaps the greatest source of stress in the woman's or girl's brain can be the fear of losing intimate relationships and the lack of vital social support that might ensue.

A girl's increasing stress and anxiety response at puberty may even be related to the formation of cliques and clubs. In fact, the formation of cliques may be the result of her stress response. Until recently, it was assumed that all human beings reacted to stress with the "fight or flight" response, a behavior described by W. B. Cannon in 1932. A person under stress or threat, the theory goes, will attack the source of that threat if there's any reasonable chance of winning; otherwise, an individual will flee from a threatening situation. "Fight or flight" behavior, however, may not be characteristic of all humans. The UCLA psychology professor Shelley Taylor argues, in fact, that this is more likely to be the *male* response to threat and stress.

Both sexes, to be sure, experience a powerful flood of neurochemicals and hormones when they come under acute stress, preparing them to meet the demands of an imminent threat. And that cascade can make males spring into action—their aggression pathways are more direct than females'. But fighting may not have been as evolutionarily adaptive for females as it was for males because females have less chance of physically defeating the larger males, and even if they are matched in strength with their opponents, turning to fight could mean leaving a helpless child alone and vulnerable. In the female brain, the circuit for aggression is more closely linked to cognitive, emotional, and verbal functions than is the male aggression pathway, which is more connected to brain areas for physical action.

As for flight, females are generally less able to run when they're pregnant, nursing, or caring for a vulnerable child. Research has found that female mammals under stress rarely abandon their infants once they've formed maternal bonds. As a result, females appear to have some stress responses in addition to "fight or flight" that allow them to protect themselves and their dependent children. One such response may be reliance on social ties. Females in a bonded social group are more likely to come to one another's aid in a threatening or stressful situation. Members of a group can alert one another to conflict ahead of time, enabling them to move away from potential danger and continue safely tending dependent children. This pattern of behavior is termed "tend and befriend," and it may be a particularly female strategy. Tending involves nurturant activities that promote safety and reduce distress for the self and offspring; befriending is the creation and maintenance of social networks that may aid in this process.

Remember, our modern female brain still has the ancient circuitry of our most successful foremothers. Early in mammalian evolution, females may well have formed social networks for support when threatened by males, as studies of some nonhuman primates indicate. In certain species of monkeys, for example, if a male is overly aggressive to a female, the other females in her group will come and face the male down, standing shoulder to shoulder, chasing him away with threatening cries. These female networks provide other types of protection and support, too. Many species of female primates will watch and care for one another's infants, share information about where to find food, and model maternal behavior for younger females. The UCLA anthropologist Joan Silk found a direct link between the degree of social connectedness of female baboons and their reproductive success. Her sixteen-year study showed that mothers who were the most socially connected had the greatest number of surviving infants, increasing their success at passing on their genes.

Teen girls begin automatically building and practicing these friendship connections during their intimate talks in the school's bathrooms. Biologically, they are reaching optimal fertility. The Stone Age brains within them are flooded with neurochemicals telling them to connect with other women so that they can help protect the young. The primitive brain is saying, "Lose that bond, and both you and your offspring are toast." That's a powerful message. No wonder girls find it unbearably hard to cope with feelings of being left out.

THE BRAIN MARCHES TO THE BEAT OF ESTROGEN'S DRUM

By the time Shana was ten years old, it was harder for Lauren to wake her up for school. Shana started sleeping until noon on weekends. Lauren was sure this sleep pattern reflected Shana's bad habits—she waited until the last minute to finish big projects, and she liked to stay up watching television. Shana was beginning to feel depressed because her mom was calling her a lazy bum all the time, but Shana couldn't see why. She was tired and wanted to sleep. Mother and daughter were locked in battle when I first saw them.

In reality, the sleep cells in Shana's brain had been reset at puberty by her ovarian estrogen surges. Estrogen affects practically everything that a teen girl experiences, including responsivity to light and the daily light-dark cycle. Estrogen receptors get activated in the brain's twenty-four-hour clock cells in the suprachiasmatic nucleus. These clusters of cells orchestrate the daily, monthly, and annual rhythms of the body, such as those of hormones, body temperature, sleep, and mood. Estrogen even directly influences the brain cells that control breathing. It turns on the uniquely female sleep cycle as well as her growth hormone.

By puberty, estrogen sets the timing of everything in the female brain—the female and male brains end up marching to different drummers.

At around age eight to ten for girls—and a year or more later for boys—the brain's sleep clock begins to change its settings, leading to later bedtimes, later wakeup times, and more sleeping time overall. One study showed that at age nine boys' and girls' brains had exactly the same brain waves during sleep. By age twelve girls had a 37 percent shift in their brain waves during sleep compared to boys. The scientists concluded that this indicated that girls' brains mature faster. The pruning of excess synapses in teen girls' brains starts earlier than it does in boys, thus moving them more quickly toward maturation of all their brain circuits. The female brain, on average, matures two to three years earlier than the male brain. A similar condition develops in boys' brains a few years later, but their sleep phase is pushed even an hour later than girls' by the age of fourteen. And this is just the beginning of being out of sync with the opposite sex. Females' tendency to go to sleep and wake up a bit earlier than do males is a difference that will last until after menopause.

I saw Shana and her mother many times over the years. Things became even rockier as Shana got several years into the new rhythm that estrogen was establishing in her brain. It was day twenty-six of her cycle, and Shana wasn't just screaming. She was shrieking. "I am going to the beach tomorrow and there's nothing you can do about it. Just try to stop me."

"No, Shana," Lauren responded, "you're not going with that group of kids. I told you I don't like the fact that they throw around so much money, and I'm pretty sure they're into drugs."

"You don't know what you're talking about. You're a stupid old prude who just doesn't have a life. You never had one. You were ugly and boring and a goody-two-shoes kid. You wouldn't know cool if it smacked you in the face. You can't stand it that I'm smarter than you and cooler than you, and you just want to keep me down. You're a fucking asshole!"

Lauren lost it. For the first time in her life, she slapped her daughter.

The most obvious cycle controlled by estrogen is the menstrual cycle. The first day a young girl gets her period can be exciting and bewildering. It is a moment to celebrate, not in a New Age, hippie sense but because each month the menstrual cycle refreshes and recharges certain parts of a girl's brain. Estrogen acts as a fertilizer on cells— exciting her brain as well as making a girl more socially relaxed during the first two weeks. There's a 25 percent growth of connections in the hippocampus during weeks one and two (the estrogen phase), and that makes the brain a little bit sharper. It functions a little better. You're clearer and you remember more. You think more quickly and more agilely. Then at ovulation, around day fourteen, progesterone starts squirting out of the ovaries and reversing the actions of estrogen, acting more like weed killer on those new connections in the hippocampus. During the last two weeks of the cycle, progesterone causes the brain to become first more sedated and gradually more irritable, less focused, and then a little slower. This may be one of the pivotal reasons for the change in stress sensitivity during the second half of the menstrual cycle. The extra connections built during the weeks that estrogen is on the rise are being reversed by progesterone in the last two weeks.

In the last few days of the menstrual cycle, when progesterone collapses, this calming effect is abruptly withdrawn, leaving the brain momentarily upset, stressed, and irritable. This is where Shana was when she screamed at her mother. Many women say they cry more easily and often feel out of sorts, stressed, aggressive, negative, hostile, or even hopeless and depressed right before their periods begin. In my clinic we call them the "crying over dog food commercials" days, because even silly, sentimental things can trigger a tearful response during this short time. At first this abrupt mood change takes girls like Shana by surprise. Teens think that all they need to know about the menstrual cycle is to remember their Tampax and take Advil or Aleve for the cramps on the day the blood flow starts. The idea that even when they're not bleeding there could be brain effects from their cycling hormones takes some getting used to. By adulthood, they know how to handle it. Most women know that, in weeks three and four, angry impulses fall under the two-day rule. They'll wait two days and see if they still want to act on them.

It took another few days for Shana to realize she should not have spoken to her mother the way she did. And as her progesterone cycled out and her estrogen came back up, her irritability began to wane. Connections were once again sprouting in the hippocampus, and her brain gears were greased and working to full capacity. Pretty soon she was surprising everyone with her wisecracks and smart-alecky remarks, and they were getting her into a bit of trouble—the boys just couldn't keep up at times, and she was riding the edge with the girls. Brain performance in some females can fluctuate with the hormonal changes of the menstrual cycle. One of the most estrogen-sensitive parts of the brain—the hippocampus—is a major relay station for processing memories for words. This may be one biological reason behind women's increased verbal performance during the highest estrogen week—week two—of their cycles. I often joke with my female grad students that they should take their oral exams on day twelve of their cycles, which is the peak of their verbal performance. Maybe the same should go for teen girls and the SATs—or for wives wanting to win a fight with their husbands.

WHY THE TEEN GIRL BRAIN FREAKS

Think about it. Your brain has been pretty stable. You've had a steady flow—or lack—of hormones for your entire life. One day you're having tea parties with Mommy, the next day you're calling her an asshole. And, as a teen girl, the last thing you want to do is create conflict. You used to feel like a nice girl, and now, out of nowhere, it's as though you can't rely on that personality anymore. Everything you thought you knew about yourself has suddenly come undone. It's a huge gash in a girl's self-esteem, but it's a pretty simple chemical reaction, even for an adult woman. It makes a difference if you know what's going on.

The trouble for some women is caused by estrogen and progesterone withdrawal in the brain, which happens in the fourth week of the cycle. The hormones bottom out precipitously, and the brain begins yearning for their calming effects. When it doesn't get them, the brain becomes irritated, so irritated that it's on the same spectrum of discomfort as a seizure. This is true in a small percentage of women, to be sure, but it's not fun. So stress and emotional reactivity can increase dramatically the few days before the onset of bleeding. At the National Institute of Mental Health in Bethesda, Maryland, David Rubinow and colleagues have been studying menstrual mood changes. They've now found direct evidence that the hormone fluctuations during the menstrual cycle increase brain circuit excitability, as measured by the startle reflex, which most of us think of as being jumpy. It is also related to the stress response. This helps explain why women often feel more irritable during maximal hormone withdrawal.

Although 80 percent of women are only mildly affected by these monthly hormone changes, about 10 percent say they get extremely edgy and easily upset. Females whose ovaries make the most estrogen and progesterone are more resistant to stress because they have more serotonin (a chemical that makes you feel at ease) cells in their brains. Those women with the least estrogen and progesterone are more sensitive to stress and have fewer serotonin brain cells. For these most stress-sensitive individuals, the final days before their periods start can be hell on earth. Hostility, hopeless feelings of depression, plans for suicide, panic attacks, fear, and uncontrollable bouts of crying and rage can plague them. Hormone and serotonin changes can result in a malfunction in the brain's seat of judgment (the prefrontal cortex), and dramatic, uncontrolled emotions can push through more easily from the primitive parts of the brain.

Shana was in this category. During the week or two before her period, she was constantly in trouble for talking out of turn and being disruptive in class. She was obnoxious and aggressive one minute, bursting into tears the next. Pretty soon, her moods turned wild, and she began to intimidate her parents, peers, and teachers. Repeated meetings with the principal and school counselor did nothing to curb her outbursts, and when her parents finally sent her to a pediatrician, he too was baffled by her extreme behavior. It was a female teacher who noticed that Shana's behavior was at its worst during two weeks of each month. The rest of the time she was like her old self—or more like a typical teen—sometimes moody and oversensitive but mostly cooperative. On a hunch, the teacher called me at the clinic to suggest that Shana had bad PMS.

Shana's mood and personality swings, while extreme, were no surprise. In twenty years of practice in psychiatry and women's health, I've seen hundreds of girls and women with similar problems. Most blame themselves for their flare-ups of bad behavior. Some have been in psychotherapy for years trying to get to the bottom of their recurring sadness or anger. Many have been regularly accused of substance abuse, bad attitudes, and bad intentions. Most of these assumptions are unjust, and all of them completely miss the point.

These adolescent girls and adult women have regular, dramatic shifts in their moods and behavior because, in fact, the very structure of their brains is changing, from day to day and from week to week. The medical name for an extreme emotional reaction during the weeks before the period, triggered by ovarian estrogen and progesterone hormones, is premenstrual dysphoric disorder (PMDD). Women who have committed crimes while suffering from PMDD have successfully used it as a defense in France and England by establishing temporary insanity. Other common conditions—such as menstrual migraine—are also caused by increased brain circuit excitability and decreased calming right before the period starts. Researchers at the National Institute of Mental Health found that the emotion and mood changes these women experience during the menstrual cycle disappear when the ovaries are blocked from producing fluctuating hormones. It may be, they conclude, that women with PMDD are in some sense "allergic" or hypersensitive to fluctuations in estrogen and progesterone during the cycle. Fifty years ago, one successful treatment for PMDD was removing the ovaries surgically. At the time, this was the only way to remove the hormone fluctuation.

Instead of removing Shana's ovaries, I gave her a hormone to take every day—the continuous birth control pill—to keep her estrogen and progesterone at moderately high but constant levels and prevent her ovaries from sending out the big fluctuations of hormones that were upsetting her brain. With her estrogen and progesterone at constant levels, her brain was kept calmer and her serotonin levels stabilized. For some girls I add a medication such as Zoloft—a so-called SSRI (selective serotonin reuptake inhibitor)—which can further stabilize and improve the brain's serotonin level, in other words, improve one's mood and sense of well-being. The following month her teacher called me to report that Shana was back to her good old self again— cheerful and getting good grades.

RISK TAKING AND AGGRESSION IN TEEN GIRLS

The day Shana screamed that she wanted to go to the beach, Lauren had been concerned about her daughter's boyfriend, Jeff. Jeff was from a very wealthy and permissive family, and at fifteen, Shana had already had sex with him. Jeff's parents allowed them to do it in their house, a fact Shana had kept hidden from her parents until she had a pregnancy scare. Since Jeff wasn't going away, Lauren decided it was best to get to know him. And the more she did, the more she liked him. Jeff was lavishing Shana with gifts (something Lauren wasn't thrilled about, but she didn't want to hurt his feelings), and Shana was happy when he was around. She made deals with her parents: "Come on, Mom, I'm really stressed out, and if he comes over for an hour I'll feel better. I promise to finish my work after he leaves." She often snuck him back in, and the two were thick as thieves.

Shana had been seeing Jeff for eight months. The day after she told her mom how much she loved him, Shana showed up at home after school with Mike, a boy she had sworn was just a friend. When Lauren went up to check on them, the door was closed. When she opened it, they were, as she put it, "sucking face." Since she had allowed Shana to be sexually involved with Jeff, Lauren didn't know what to do. It was clear that Shana's impulses were getting out of control.

A girl's emotional centers become highly responsive at puberty. Her brain's emotion- and impulse-control system—the prefrontal cortex—has sprouted many more cells by the age of twelve but the connections are still thin and immature. As a result, a teenage girl's mood changes, resulting in part from the increased emotional impulses blasting in from the amygdala, are more rapid and dramatic. Her prefrontal cortex is like an old dial-up modem receiving signals from broadband. It can't handle the increased traffic from the amygdala, and it often becomes overwhelmed. Teenagers, therefore, often cling to an idea and run with it, not stopping to consider the consequences. They become resentful of any authority that wants to head off their impulses.

My patient Joan, for example, remained in upstate New York the summer after she graduated from boarding school there. An honors student, she had been involved with a local guy who didn't graduate high school, had been in juvenile detention, and at age sixteen had fathered a child. She ran around with him all summer, and when it came time to leave for college, she thought twice about it. She wanted to stay with him. When her parents threatened to come up, take the car, and drag her off to college, she ran away with her boyfriend. She did come to her senses and go to college, but it was a long time before she spoke civilly to her parents again. It's tough for the teen brain to come up with good judgment in these situations.

Remember Romeo and Juliet? If only the two lovers had known that their brain circuits were under major reconstruction. If only they'd known that their sex hormones were causing brain cells to grow and sprout extensions, and that it would take several years to form structurally sound connections once those extensions were plugged into the correct outlets in mature prefrontal cortexes. Juliet's brain would have matured two to three years earlier than Romeo's, though—so she may have come to her senses sooner than he. These unfinished— unmyelinated— extension cords, most prominent in the connections from the emotion center of the amygdala to the emotional control center of the prefrontal cortex, need to be coated with a substance called myelin that allows for fast conduction before they can function reliably under stress. This may not happen until the late teens or early adult years. Without the fast connection to the prefrontal cortex, big downloads of emotional impulses often result in immediate, raw behaviors and circuit overload.

When it is upset by an unwanted parental restriction such as, "We know you were drinking at the party, and you're too involved with boys, and your grades are going down, so you are grounded," the teen girl amygdala may not respond with much more than "I hate you." But watch out for the subtle signs of rebellion that can ensue. She'll find another way to undermine you.

Karen, a former patient of mine who is now a tenured professor of biochemistry, told me a story that illustrates this teen reality. She grew up in a small town in Washington State, where many students dropped out of high school to work for the lumber companies in the area. Her girlfriends got jobs as cooks or secretaries in the lumber camps, or got married and almost immediately became pregnant. By the time she was a sophomore in high school, Karen was desperate to get away from home. She was determined to go to college, a radical idea in a town where only the teachers, the doctor, and the librarian were college graduates. Her parents accused her of living in a fantasy world. They didn't have the money to send her, and what did she think she could do with a college degree when she would likely get "knocked up" by the time she was twenty?

Their scorn made Karen even more determined to find a way out. At eighteen, she wanted to stay in school and graduate. But she was old enough to get a job as a go-go dancer in one of the local bars that serviced the lumberjacks who came into town to spend their paychecks. She moved in with her boyfriend and worked nights at the bar. Too young to go topless, she still managed to earn twenty-dollar tips, which the customers would stick in her G-string.

Not exactly the typical line of work for a future biochemistry professor. But Karen earned enough money to pay for her first semester at college, and after that, her grades were rewarded with a full scholarship. Now that Karen is herself the mother of three teenagers, two girls and a boy, she tries to imagine how she would react if her eighteen-year-old daughter announced that she had just gotten herself a job as a pole dancer in a bar. She herself had avoided any dangerous incidents, but her go-go gig could have turned out otherwise.

CHANGING HORMONAL CONDITIONS in girls' brains through the menstrual cycle add even more volatility to the mix. If estrogen and progesterone simply increased during the teen years and remained at that new, higher level, the female brain would permanently readjust. But, as we have seen, these hormones come in waves. Given the fact that the teen brain is undergoing major changes, especially in areas that are particularly sensitive to shifts in hormones, puberty can be an outrageously impulsive time for many girls. Under no stress on a good week of the menstrual cycle, the teen girl's prefrontal cortex may function normally. At those times she may have good judgment and appropriate behavior. But some stress—like a disappointment or a bad grade—on a PMS day may derail the prefrontal cortex, causing an exaggerated emotional response and out-of-control behavior, such as yelling and slamming doors, what at my house we call a meltdown. Testosterone surges in teen boys may have similar brain effects, but these have not been studied yet. Hormone surges at this age can make a mild stress or a seemingly small event feel like a catastrophe.

Calming down the fired up teen girl amygdala can prove difficult. Many girls turn to drugs, alcohol, and food (either they'll stop eating or they'll binge) when they're under stress. As a parent of teens, you have the job of ignoring much of what they say. Don't take any impulsive or emotional tirades seriously. Stay calm. Teens state their intentions—and feel them—with such passion—that you can be persuaded in spite of yourself. Just remember, your teen daughter's impulse-control circuits can't handle the input. Like it or not, you must provide the control while her brain cannot. Even though Joan hated her parents for threatening to come and take her car away, "they did the right thing," she told me years later. It was their job to exercise the good judgment that she lacked at that time.

DEPRESSION

It wasn't long before even Mike began to realize Shana's impulses were out of control. If she could turn on a dime about Jeff, she could change her mind about him, too, and he decided to break it off. A few of Shana's friends were also mad at her over how she'd treated Jeff, and she was becoming isolated. Up until then, Shana had been doing well. She was writing for the school newspaper, was becoming serious about sculpting, and was going to have a good pick of colleges. Her teachers loved her creativity and spark. But when Mike broke it off, everything changed. Shana lost a ton of weight. She stopped doing well in school. She let the school paper down by not writing stories that were assigned. She couldn't concentrate or do her homework, couldn't sleep, was obsessed with her weight and appearance, and couldn't get her brain to stop thinking about *him*. I could see a few scratches on her arm and realized she was cutting. I was quite alarmed, since this is the period when the female-to-male ratio for depression doubles.

Boys and girls have the same risk of depression, before the hormones of puberty. But by age fifteen, girls are twice as likely to suffer from depression. Genetics may also play a role in female depression. In certain families with high depression rates, for example, researchers have found a mutation in a gene called CREB1 that puts teenage females—but not males—at higher risk for clinical depression. Shana's mother and grandmother had had serious depressions in their teens, and a female cousin had committed suicide. These facts put her at serious risk. Shana had a true clinical depression. I started her on Zoloft, stayed in close contact, and did weekly cognitive therapy. Within four to six weeks she was able to concentrate again, take her final exams, and stop obsessing over both Mike and her weight.

THE BIOLOGY OF MEAN GIRLS

Hormonal surges can turn nice girls mean at the drop of an egg, and so can sexual competition, which is strong—and pivotal—among teenage girls. This competition, however, plays out with a different set of rules than does that among teen boys. Girls are driven to gather in cliques, but there is another side, in which these cliques are at war. Teen girls, we know, can be devastatingly mean. When females are competing with other females, they often use more subtle tools, such as spreading rumors to undermine a rival. This way, they can cover their tracks—"I wasn't trying to be mean. I'm sorry." Such tactics lessen the risk of destroying the bond that the teen girl brain sees as essential to survival. But also essential to survival is sexual competition.

I can remember when I was in seventh grade, there was one girl who was beautiful, and the other girls were very jealous because she got so much attention from the boys. She was also shy, so others assumed she was a snob. One day the not as pretty girl who sat directly behind her in a class took a wad of bubble gum out of her mouth and stuck it in the pretty girl's hair. Unknowingly, the pretty girl began to twist the gum into such a mess that the only way to get it out was to cut off her seductive locks. The queen of mean who put the gum in this girl's hair felt triumphant. Her biological imperative to compete for sexual attractiveness had a momentary victory.

The hormones usually associated with aggression in both males and females are androgens. They begin to rise early in puberty and continue until they peak at age nineteen in females and twenty-one in males. The three main androgens that women make are testosterone, DHEA, and androstenedione (andro-steen-DIE-own). In a study at the University of Utah, the most in-your-face aggressive teenage girls were found to have high levels of the androgen androstenedione. Acne is a good clue that your teen's androgen levels are high. Girls with high levels of testosterone and DHEA also tend to have sexual intercourse earlier. By the time I saw Shana at age fifteen, she not only had acne and fully developed breasts but had been having sex for the past year.

Aggressive impulses can fluctuate with the hormones of the menstrual cycle. During some weeks of the cycle, the teen girl will be more interested in social connection. During other weeks, she'll be more interested in power—over boys and other girls. This association implies that the higher amounts of androgens made by the ovaries during weeks two and three increase aggression levels in women and teens. Less empathy, social connection, and affiliation have been associated with higher androgen levels in both sexes. We can't know for sure, but Shana's higher androgen levels on certain weeks of her cycle may have been triggering her aggressive outbursts.

Not only is aggression reduced when androgen levels are low but sex drive is decreased, too. Teens taking oral contraceptives have reduced aggression and sex drive because the contraceptive suppresses the ovaries, so they make less androgen. Although both men and women make testosterone, men make more than ten times as much—meaning that their sex drive is much greater than women's. Scientists know that it is probably not just androgens that increase aggressive spirit and ambition in women but estrogen, too. In the same study at the University of Utah, women who were the most outspoken and had the highest self-regard also had the highest levels of estrogen, testosterone, and androstenedione. They also ranked themselves above how their peers ranked them. And these young women were routinely rated by others as the most boastful.

Of course, a hormone alone does not cause a behavior. Hormones merely raise the likelihood that under certain circumstances a behavior will occur. And just as there is no one seat of aggression in the brain, there is no one hormone of aggression. But achieving success and attaining power in the world requires some aggression for both sexes. These hormones change teens' reality and perceptions of themselves as sexual, assertive, and independent beings in the world.

During the teen years a girl's brain circuits go through massive growth and pruning. It's as if she is given a whole new set of extension cords and needs to figure out which one to plug into which outlet. The full power of her female brain circuits can now start to be manifested. And where will they push her? Right into the arms of a man.

THREE

Love and Trust

MELISSA, A BRASSY San Francisco film producer, really wanted to fall in love. Her career was finally chugging along at a steady pace, and at age thirty-two, she was ready to move into the next phase of her life. She now wanted a family and the continuity of a relationship with a man who would stick by her for more than a few sexually charged months. The only problem was that she couldn't seem to connect with the right one. She would go on countless dates through setups, or with men she met on the Internet, but none was touching off the flurry of butterflies in her stomach or that intense, irrational need to be around him all the time.

One night her best friend, Leslie, called and asked Melissa to go salsa dancing. But Melissa wasn't in the mood. She wanted to stay home, relax, and watch TV, but Leslie was relentless, so Melissa acquiesced. She tousled her curly hair to look sexy, put on a swirly skirt, her new red suede heels, and bee sting red lipstick, which made her mouth pop out. She grabbed a taxi over to the dance club.

Leslie was already inside drinking a margarita when Melissa arrived. As they were getting loose to hit the dance floor, Melissa saw a tall, handsome man with a sculpted face, olive skin, and a shock of nearly black hair across the room. "Wow, he's gorgeous," she said.

She turned back to Leslie and whispered for her to glance over at the man, but it was too late. He was already walking toward them. Melissa was locked in gaze with this stranger. A wave of energy shot up her back. It was the feeling she hadn't experienced in all the months of her bad dates. There was something vaguely familiar about him. "Hmm, who *is* that?" she whispered under her breath to Leslie, as her brain's cortex scanned her memory banks. No match was found, but all her attention circuits were now on "mating alert status." Is he here alone or with someone? she wondered. She looked around for the one of the gorgeous women who always seem to be attached to these perfect-looking guys but saw no one. And he was still walking toward her.

The closer he got, the more unfocused Melissa became on her friend's story. She grabbed her drink tightly. Her eyes and attention were riveted on him, taking in every detail—his leather Armani shoes, his sexy black cords, and no wedding ring on his left finger. Everything else dropped into the background as her brain honed to make contact. She felt like she was falling in love. The mating impulse had taken over.

"Hi, I'm Rob," he said, leaning against the bar nervously. His voice was pure velvet. "Have we met before?" Melissa was unable to hear his words. She could only bask in the feel of him, his earthy smell, and his devilish green eyes.

The dance of romance had begun, and its choreographer was not her friend or a matchmaker. It was the biology of Melissa's brain. We know that the symmetry of physiques and faces that entrance us, the moves that seduce us, and the heart-pounding passion of attraction are all hardwired into our brains' love drive by evolution. Short- and long-term "chemistry" between two people may seem accidental, but the reality is that our brains are preprogrammed to know better. They subtly but firmly steer us toward partners who can boost our odds in the sweepstakes of human reproduction.

Melissa's brain is beginning to imprint Rob. Her hormones are surging. As he tells her that he is a marketing consultant who lives in a loft in Potrero Hill and musters up the nerve to ask her to dance, her brain, faster than a supercomputer, calculates the qualities that might put him in the running as a mating partner. Already some green light is flashing that he's a good one, and *wham*, hot, knee-buckling waves of attraction and desire are flooding her body with a heady rush of dopamine—sparking euphoria and excitement. Her brain has also ordered her a shot of testosterone, the hormone that stokes sexual desire.

As Rob speaks, he is also sizing up Melissa at a closer view. If his calculations come out positive, he'll get a neurochemical jolt, too, prodding him to try to hook up with her. With their love circuits mutually revved up, the two move onto the dance floor and spend the next few hours locked in sweaty salsa rhythms. At 2:00 A.M., the music slows down and the club begins to empty. Leslie has gone home hours earlier. Standing on the corner, Melissa says that she has to go and flirtatiously turns on her high heels. "Wait," Rob says. "I don't have your number. I want to see you again." "Google me and you'll find me," she replies, smiling and jumping into a cab. Now the chase begins.

For men and women, the initial calculations about romance are unconscious, and they're very different. In short-term couplings, for example, men are chasers and women are choosers. That's not sex stereotyping. It's our inheritance from ancestors who learned, over millions of years, how to propagate their genes. As Darwin noted, males of all species are made for wooing females, and females typically choose among their suitors. This is the brain architecture of love, engineered by the reproductive winners in evolution. Even the shapes, faces, smells, and ages of the mates we choose are influenced by patterns set millennia ago.

The truth is, we're much more predictable than we think. Over the course of our evolution as a species, our brains have learned how to spot the healthiest mates, those most likely to give us children, and those whose resources and commitment can help our offspring survive. The lessons that early men and women learned are deeply encoded in our modern brains as neurological love circuits. They are present from the moment we're born and activated at puberty by fast-acting cocktails of neurochemicals.

It's an elegant system. Our brains size up a potential partner, and if he fits our ancestral wish list, we get a jolt of chemicals that dizzy us with a rush of laser-focused attraction. Call it love or infatuation. It's the first step down an ancient pair-bonding path. The gates have opened to the courtship-mating-parenting brain program. Melissa may not have wanted to meet anyone that night, but her brain had other plans that are deep and primitive. When it saw Rob across the room, a signal went off for mating and long-term attachment, and she was lucky that his brain felt the same way. Each of them will come up against anxiety, threats, and mind-numbing joys, over which they have little control because now biology is building their future together.

MIND-SET ON MATING

As Melissa struts along the city streets, sips her latte, or cruises the Internet for potential dates while she's waiting for Rob to locate her number on her website—she did tell him the name of her latest film, so if he is smart, he'll find her—it's not easy to believe that what's inside her cranium is a Stone Age brain. But that's the case, according to scientists who study the mate-attraction engineering of the human mind. We spent more than 99 percent of the millions of years it took human beings to evolve living in primitive conditions. As a result, the theory goes, our brains developed to solve the kinds of problems that those early human ancestors encountered. The most important challenge they faced was reproduction. It wasn't just a matter of having children. It was making sure those children lived long enough to propagate their genes. Early people whose mating choices produced more surviving offspring succeeded in passing their genes on. Their specific brain systems for courtship attraction were more successful. Ancestors who made the wrong reproductive moves left no imprint on the future of the species. As a result, the brain wiring of the best Stone Age reproducers became the standard-issue circuitry of modern humans. This courtship circuitry is what is commonly known as "falling in love." We may think we're a lot more sophisticated than Fred or Wilma Flintstone, but our basic mental outlook and equipment are the same.

That our mental instincts haven't changed in millions of years may explain why women, worldwide, look for the same ideal qualities in a long-term mate, according to the evolutionary psychologist David Buss. For over five years, Buss studied the mate preferences of more than ten thousand individuals in thirty-seven cultures around the world—from West Germans and Taiwanese to Mbuti Pygmies and Aleut Eskimos. He discovered that, in every culture, women are less concerned with a potential husband's visual appeal and more interested in his material resources and social status. Rob had told Melissa he was a marketing consultant—they were a dime a dozen in San Francisco, and Melissa had seen more than a few go out of business. She didn't realize that this thought was making it hard for her to figure out if Rob was Mr. Right or Mr. Right Now.

Buss's findings may be uncomfortable at a time when many females are achieving at high levels and are proud of their social and financial independence. Nevertheless, he found that, in all thirty-seven cultures, females value these qualities in a mate much more than males do, regardless of the females' own assets and earning capacity. Melissa may be an independent economic unit, but she wants her partner to provide, too. Female bowerbirds share this preference by choosing to mate with the male who has built the most beautiful nest. My husband jokes that he's like a male bowerbird, since he built a beautiful house several years before we met, and it was ready and waiting for me. Women, researchers have found, also look for mates who are, on average, at least four inches taller and three and a half years older. These female mate preferences are universal. As a result, scientists conclude, they're part of the inherited architecture of the female brain's mate-choice system—and are presumed to serve a purpose.

According to Robert Trivers, a pioneering evolutionary biologist at Rutgers University, choosing a mate based on these attributes is a savvy investment strategy. Human females have a limited number of eggs and invest far more in bearing and raising children than males do, so it pays for women to be extra careful with their "family jewels." This is why Melissa didn't jump into bed with Rob on the first night, even though the dopamine and testosterone surging through her brain's attraction circuits made him hard to resist. It's also why she kept a number of other guys on her dance card. While a man can impregnate a woman with one act of intercourse and walk away, a woman is left with nine months of pregnancy, the perils of childbirth, months of breast feeding, and the daunting task of trying to ensure that child's survival. Female ancestors who faced these challenges alone were likely to have been less successful in propagating their genes. Though single motherhood has become fashionable among some sets of modern women, it remains to be seen how well this model will succeed. Even today, in some primitive cultures, the presence of a father triples children's survival rate. As a result, the safest bet for females is to partner long-term with males who are likely to stick around, protect them and their children, and improve their access to food, shelter, and other resources.

Melissa was smart to take her time and make sure Rob was a good catch. Her dream was a husband whom she loved, and who loved and worshiped her back. Her worst fear was a man who might be unfaithful, the way her father was to her mother. After the night at the dance club, she got a number of positive clues. Rob was taller, older, and appeared financially comfortable. In the grand, Stone Age scheme of things, he fit the bill, but it still wasn't clear whether he was the long-term type.

CHEMICAL ATTRACTION

If Melissa's ancient brain circuitry was scanning for assets and protection, what was Rob's brain looking for in a long-term partner? According to Buss and other scientists, something completely different. Worldwide, men prefer physically attractive wives, between ages twenty and forty, who are an average of two and a half years younger than they are. They also want potential long-term mates to have clear skin, bright eyes, full lips, shiny hair, and curvy, hourglass figures. The fact that these mate preferences hold true in every culture indicates that they're part of men's hardwired inheritance from their ancient forefathers. It wasn't just that Rob had a thing for girls with shiny curls. Melissa's hair triggered his ancient attraction wiring.

Why would these particular criteria top men's lists? From a practical perspective, all of these traits, superficial as they may seem, are strong visual markers of fertility. Whether or not men know it consciously, their brains know that female fertility offers them the biggest reproductive payoff for *their* investment. With tens of millions of sperm, men are capable of producing an almost unlimited number of offspring as long as they can find enough fertile females to have sex with. As a result, their key task is to pair up with women who are likely to be fertile and reproduce. Pairing with infertile women would be a waste of their genetic futures. So, over millions of years, male brain wiring evolved to scan women for quick visual clues to their fertility. Age, of course, is one important factor; health is another. A high activity level, youthful gait, symmetrical physical features, smooth skin, lustrous hair, and lips plumped by estrogen are easily observable signs of age, fertility, and health. So it's no wonder women are reaching for the plumping effects of collagen injections and the wrinkle smoothing of Botox.

Shape, too, is a remarkably good indicator of fertility—breast implants notwithstanding. Before puberty, males and females have very similar body shapes and waist-to-hip ratios. Once the reproductive hormones kick in, however, healthy females develop curvier shapes, with waists that are about one-third narrower than their hips. Women with that body type have more estrogen and become pregnant more easily and at a younger age than those with waists that are closer in size to their hips. A thin waist also gives an instant clue to a woman's reproductive availability, since pregnancy radically alters her silhouette. Social reputation is often a factor in male assessment, since the most reproductively successful males also need to pick women who will mate only with them. Men want to ensure their paternity but also to be able to count on a woman's mothering skills to make sure that their offspring thrive. If Melissa had immediately gone to bed with Rob or showed off to him about all the guys she had had, his Stone Age brain might have judged that she would be unfaithful or had a bad reputation. That she was affectionate on the dance floor and went home at a proper hour in a taxi showed him she was a high-quality lady with whom to mate long-term.

CALCULATING POTENTIAL DANGER

Rob left a message on her machine, and Melissa waited a few days before calling him back. And although they had kissed on the first date, she had no plans of going to bed with him until she knew more about him. He was incredibly funny and charming, and seemed to have his life in order, but she needed to be sure on a gut level that she could trust him. The brain's anxiety circuits usually fire around strangers—her amygdala's fear circuits were still turned on full force. A natural cautiousness toward strangers is part of the brain wiring of both males and females, but women in particular give early, careful scrutiny to a man's likely level of commitment when looking for a mate.

Seduction and abandonment by males is an old ruse, going back to the beginning of our species; one study found that young college males admitted to depicting themselves as kinder, more sincere, and more trustworthy than they really are. Some anthropologists speculate that natural selection favored men who were good at deceiving women and getting them to agree to have sex. Females, as a result, had to get even better at spotting male lies and exaggerations—and the female brain is now well-adapted to this task. A study by the Stanford University psychologist Eleanor Maccoby showed, for example, that girls learn to tell the difference between reality and fairy tales or "just-pretend" play earlier than boys. By adulthood, modern females have fine-tuned their superior ability to read emotional nuance in tone of voice, eye gaze, and facial expressions.

As a result of this extra cautiousness, the typical female brain isn't as ready to admit to being overwhelmed by infatuation or the sheer excitement of sexual behavior as is the male. Women do reach the same or a higher romantic end point, but they're often slower to confess to being in love and more careful than males in the beginning weeks and months of a relationship. Male brains have a different neurological love wiring. Brain-imaging studies of women in love show more activity in many more areas, especially gut feelings, attention, and memory circuits, while men in love show more activity in high-level visual processing areas. These heightened visual connections may also explain why men tend to fall in love "at first sight" more easily than women.

Once a person is in love, the cautious, critical-thinking pathways in the brain shut down. Evolution may have made these in-love brain circuits to ensure we find a mate and then focus in exclusively on that one person, according to Helen Fisher, an anthropologist at Rutgers University. Not thinking too critically about the loved one's faults would aid this process. In her study on being in love, more women than men said that their beloveds' faults don't matter much to them, and women scored higher on the test of passionate love.

THE BRAIN IN LOVE

Melissa and Rob were talking on the phone almost every night. Every Saturday they would meet in the park to take Rob's dog for a walk, or at Melissa's apartment to watch the dailies on her latest film. Rob was feeling stable in his job and had finally stopped talking about his former girlfriend, Ruth. This waning attachment to Ruth gave Melissa a clue that she wasn't just a rebound and that he was ready to focus in on her exclusively. She had already, involuntarily, fallen in love with him but hadn't told him yet. She began warming to his physical affection, allowing her sex drive to catch up with her love drive.

Finally, after three months, Melissa and Rob fell passionately into bed after a day lying in the sun at the park totally entranced by each other. The pair was tumbling into full-blown consummated love.

Falling in love is one of the most irrational behaviors or brain states imaginable for both men and women. The brain becomes "illogical" in the throes of new romance, literally blind to the shortcomings of the lover. It is an involuntary state. Passionately being in love or so-called infatuation-love is now a documented brain state. It shares brain circuits with states of obsession, mania, intoxication, thirst, and hunger. It is not an emotion, but it does intensify or decreases other emotions. The being-in-love circuits are primarily a motivation system, which is different from the brain's sex drive area but overlaps with it. This fevered brain activity runs on hormones and neurochemicals such as dopamine, estrogen, oxytocin, and testosterone.

The brain circuits that are activated when we are in love match those of the drug addict desperately craving the next fix. The amygdala—the brain's fear-alert system—and the anterior cingulate cortex—the brain's worrying and critical thinking system—are turned way down when the love circuits are running full blast. Much the same thing happens when people take Ecstasy: the normal wariness humans have toward strangers is switched off and the love circuits are dialed up. So romantic love is a natural Ecstasy high. The classic symptoms of early love are also similar to the initial effects of drugs such as amphetamines, cocaine, and opiates like heroin, morphine, and OxyContin. These narcotics trigger the brain's reward circuit, causing chemical releases and effects similar to those of romance. In fact, there's some truth to the notion that people can become addicted to love. Romantic partners, especially in the first six months, crave the ecstatic feeling of being together and may feel helplessly dependent on each other. Studies of passionate love show this brain state lasts for roughly six to eight months. This is such an intense state that the beloved's best interest, well-being, and survival become as important as or more important than one's own.

During this early phase of love, Melissa was intensely memorizing every detail of Rob. When she had to go to L.A. for a week to show a piece of her new film project at a conference, both struggled with the separation. This was not just some fantasy; it was the pain of neurochemical withdrawal. During times of physical separation, when touching and caressing is impossible, a deep longing, almost a hunger, for the beloved can set in. Some people don't even realize how bonded or in love they are until they feel this tugging at their heartstrings when the beloved is absent. We are used to thinking of this longing as only psychological, but it's actually physical. The brain is virtually in a drug-withdrawal state. "Absence makes the heart grow fonder," your mother would say as you were moaning in pain because *he* was away. I can remember the early days of dating my husband, when I already knew he was "the one" but he didn't yet. During a brief separation he "decided" we should get married—thank goodness for dopamine and oxytocin withdrawal. His heartstrings finally got the attention of his very self-sufficient and independent male brain, as his friends and family will tell you.

During a separation, motivation for reunion can reach a fever pitch in the brain. Rob was so desperate in the middle of the week for physical contact with Melissa that he flew down to see her for a day. Once reunion takes place, all the components of the original loving bond can be reestablished by dopamine and oxytocin. Activities such as caressing, kissing, gazing, hugging, and orgasm can replenish the chemical bond of love and trust in the brain. The oxytocin-dopamine rush once again suppresses anxiety and skepticism and reinforces the love circuits in the brain.

Mothers often warn their daughters not to get too close too soon with a new boyfriend, and this advice may be wiser than they realize. The act of hugging or cuddling releases oxytocin in the brain, especially in females, and likely produces a tendency to trust the hugger. It also increases the likelihood that you will believe everything and anything he tells you. Injecting the hormone oxytocin or dopamine into the brain of a social mammal can even induce cuddling and pair bonding behavior without the usual prerequisite romantic love and sexual behavior, especially in females. And consider a Swiss experiment in which researchers gave a nasal spray containing oxytocin to one group of "investors" and compared them with another group who got a placebo nasal spray. The investors who got oxytocin offered up twice as much money as did the group who got only the placebo. The oxytocin group was more willing to trust a stranger posing as a financial adviser—feeling more secure that their investment would pay off. This study concluded that oxytocin triggers the trust circuits in the brain.

From an experiment on hugging, we also know that oxytocin is naturally released in the brain after a twenty-second hug from a partner—sealing the bond between the huggers and triggering the brain's trust circuits. So don't let a guy hug you unless you plan to trust him. Touching, gazing, positive emotional inter-action, kissing, and sexual orgasm also release oxytocin in the female brain. Such contact may just help flip the switch on the brain's romantic love circuits. Estrogen and progesterone dial up these bonding effects in the female brain, too, by increasing oxytocin and dopamine. One study has shown that on different weeks of the menstrual cycle females get more of a rewarding jolt out of their brain chemicals. These hormones then activate the brain circuits for loving, nurturing behavior while switching off the caution and aversion circuits. In other words, if high levels of oxytocin and dopamine are circulating, your judgment is toast. These hormones shut the skeptical mind down.

The drive to fall in love is always hovering in the background. Being in love, however, requires making room in your life and your brain for the beloved, actually incorporating him into your self-image via the brain's attachment and emotional memory circuits. As that process unfolds, less oxytocin and dopamine stimulation is needed to sustain the emotional bond. So spending twenty-four hours a day locked in an embrace is no longer necessary.

The basic drive for romantic attachment is hardwired in the brain. Brain development in utero, the amount of nurturing one receives in infancy, and emotional experiences all determine variations in the brain circuits for loving and trusting others. Melissa knew that her father was a philanderer, and that made her even more skeptical about falling in love and becoming attached. An individual's readiness to fall in love and then form an emotional attachment can thus be affected by the brain circuit variations caused by experience and the hormonal state of the brain. Stress in the environment can help or hinder forming an attachment. The emotional attachments and bonds we make to our early nurturing figures last a lifetime. Those early nurturing figures become part of our brain circuits via the reinforcement provided by repetitive physical and emotional caretaking experiences or their lack. Safety circuits are formed based on these experiences with nurturing, predictable, secure figures. Without those experiences, there is little or no safety circuit formation in the brain. One could still fall in love for the short term, but long-term emotional attachment may be harder to achieve and sustain.

THE MATED MIND

How does the pressing reality of the "I've gotta have him every minute of the day" feeling in the brain transmute to an "Oh, hi, you again, sweetie. How's everything?" state of mind? The hormone rushes of dopamine in the brain gradually calm down. If we had an MRI scanner to view the brain changes that occur when a woman goes from a state of early romantic love to a state of long-term coupling, we'd see the reward-pleasure circuits and the throbbing hunger-craving circuits dim down, while the attachment and bonding circuits would light up to a warm yellow glow.

We know the rapturous feelings of passionate love don't last forever—and for some, the loss of intensity can be disarming. This is how I met Melissa. After she had been involved with Rob for a year, she came to see me. She explained that, for the first five months, she and Rob had had wonderful, exciting sex every day and looked forward to each minute they spent together. Now they were living together, working at demanding jobs, and starting to talk about marriage and a family. But she had begun to "feel flat" about the relationship. Her gut feelings weren't giving her that certainty anymore. It was alarming to her that she didn't have as much interest in sex. Not that she had found or even wanted someone else. It was just that now, compared especially with the first five months of their relationship, things lacked the passion and excitement she had grown to expect. What was "wrong" with her? Was Rob the right guy? Was she normal? Could she ever be happy with him long-term if the sexual spark and intense gut feelings in their relationship were gone?

Many people, like Melissa, think the loss of the romantic high of early love is a sign that a couple's relationship is going south. In reality, however, the pair may be just moving into an, important, longer-term phase of the relationship, driven by additional neurological circuits. Scientists argue that the "attachment network" is a separate brain system—one that replaces the giddy intensity of romance with a more lasting sense of peace, calm, and connection. Now in addition to the exciting pleasure chemicals of the reward system, such as dopamine, the attachment and pair-bonding system regularly triggers the release of more of the bonding chemical oxytocin, keeping partners seeking the pleasure of each other's company. Those brain circuits for long-term commitment and bond maintenance become more active. When researchers at University College, London, scanned the brains of people who were in love relationships for an average of 2.3 years, they found that, rather than the dopamine-producing brain circuits of passionate love, other brain areas, such as those linked to critical judgment, lit up. Activity in the brain's attachment circuit is maintained and reinforced over the ensuing months and years by mutually pleasurable and positive experiences, all of which release oxytocin.

From a practical perspective, this shift from head-over-heels love to peaceful pair bonding makes sense. Caring for children, after all, would be close to impossible if mates continued to focus exclusively on each other. The downshift in love's mania and sexual intensity seems tailor-made to promote our genes' survival. It's not a sign of love grown cold, it's a sign of love moving into a new, more sustainable phase for the longer term, with bonds forged by two neurohormones, vasopressin and oxytocin.

Social attachment behavior is controlled by these neurohormones, made in the pituitary and the hypothalamus. The male brain uses vasopressin mostly for social bonding and parenting, whereas the female brain uses primarily oxytocin and estrogen. Men have many more receptors for vasopressin, while women have considerably more for oxytocin. To bond successfully with a romantic partner, males are thought to need both these neurohormones. Stimulated by testosterone and triggered by sexual orgasm, vasopressin boosts a male's energy, attention, and aggression. When men in love experience the effects of vasopressin, they have a laserlike focus on their beloved and actively track her in their minds' eyes, even when she isn't present.

Women, by contrast, are able to bond with a romantic partner once they experience the release of dopamine and oxytocin, triggered by touching and the giving and receiving of sexual pleasure. Perhaps keeping my feet warm isn't my husband's *primary* responsibility in bed, but cuddling to release oxytocin is. Over time, even the sight of a lover can cue a woman to release oxytocin.

The exceptional bonding power of oxytocin and vasopressin has been studied in great detail by Sue Carter in those furry little mammals called prairie voles, who form lifelong mating partnerships. Like humans, the voles are filled with physical passion when they first meet and spend two days indulging in virtually nonstop sex. But unlike in humans, the chemical changes in the voles' brains can be examined directly in the course of this frolicking. These studies show that sexual coupling releases large amounts of oxytocin in the female's brain and vasopressin in the male's. These two neurohormones in turn increase levels of dopamine—the pleasure chemical—which makes the voles lovestruck only for each other. Thanks to that strong neurochemical glue, the pair is mated for life.

In both males and females, oxytocin causes relaxation, fearlessness, bonding, and contentment with each other. And to maintain its effects long-term, the brain's attachment system needs repeated, almost daily activation through oxytocin stimulated by closeness and touch. Males need to be touched two to three times more frequently than females to maintain the same level of oxytocin, according to a study by the Swedish researcher Kerstin UvnäsMoberg. Without frequent touch— for example, when mates are apart—the brain's dopamine and oxytocin circuits and receptors can feel starved. Couples may not realize how much they depend on each other's physical presence until they are separated for a while; the oxytocin in their brains *keep* them coming back to each other, again and again, for pleasure, comfort, and calm. No wonder Rob flew off to L.A.

SEX, STRESS, AND THE FEMALE BRAIN

Vole studies have also highlighted attachment differences between males and females. For female prairie voles, pair bonding works best under conditions of low stress. For males, high stress works better. Researchers at the University of Maryland found that if a female prairie vole is put through a stressful situation, she won't bond with a male after she mates with him. If a male prairie vole is stressed, however, he'll quickly pair up with the first available female he finds.

In humans, too, male love circuits get an extra kick when stress levels are high. After an intense physical challenge, for instance, males will bond quickly and sexually with the first willing female they lay eyes on. This may be why military men under the stress of war often bring home brides. Women, by contrast, will rebuff advances or expressions of affection and desire when under stress. The reason may be that the stress hormone cortisol blocks oxytocin's action in the female brain, abruptly shutting off a woman's desire for sex and physical touch. For her, nine months of pregnancy followed by caring for an infant under stressful conditions makes less sense than the quick deposit of sperm does for him.

THE MONOGAMY GENE

The love lives of different subspecies of voles also offer insights into brain mechanisms for monogamy, a trait that's shared by only 5 percent of mammals. Prairie voles are champion couplers, forming monogamous, lifelong pair bonds after their marathon copulations. Montane voles, by contrast, never settle down with a single partner. The difference, scientists have discovered, is that prairie voles have what amounts to a gene for monogamy, a tiny piece of DNA that montane voles lack. As her relationship with Rob became more serious, Melissa began to worry. Was Rob a prairie vole or a montane vole?

As far as researchers know, human males represent behaviors on a spectrum from totally polygamist to totally monogamous. Scientists speculate that different genes and hormones may account for this variability. There is is a gene that codes for a particular type of vasopressin receptor in the brain. Prairie voles that carry this gene have more of the receptors in their brains than do montane voles; as a result, they're much more sensitive to the pair-bonding effects of vasopressin. When researchers injected this "missing" gene into the brains of montane voles, the normally promiscuous males instantly turned into monogamous, pair-bonded, stay-at-home dads.

Males who had a longer version of the vasopressin receptor gene showed greater monogamy and spent more time grooming and licking their pups. They also showed greater preference for their partners—even when given the chance to run off with a young, fertile, and flirtatious female. Males with the longest gene variation are the most reliable and trustworthy partners and fathers. The human gene comes in at least seventeen lengths. So the current joke among women scientists is that we should care more about the length of the vasopressin gene in our mates than about the length of anything else. Maybe someday there will be a drugstore test kit—similar to a pregnancy test—for how long this gene is, so you can be sure you're getting the best guy before you commit. Male monogamy may therefore be somewhat predetermined for each individual and passed down genetically to the next generation. It may be that devoted fathers and faithful partners are born, not made or shaped by a father's example.

Our two closest primate cousins—chimpanzees and bonobos—also have different lengths of this gene, which match their social behaviors. Chimpanzees, who have the shorter gene, live in territorially based societies controlled by males who make frequent, fatal war raids on neighboring troops. Bonobos are run by female hierarchies and seal every social interaction with a bit of sexual rubbing. They are exceptionally social and have the long version of the gene. The human version of the gene is more like the bonobo gene. It would seem that those with the longer gene are more socially responsive. For example, this gene is shorter in humans with autism—a condition of profound social deficit. Differences in partner commitment behavior may therefore be related to our individual differences in the length of this gene and in hormones.

Women, because they can have only one child every nine months, want to form faithful partnerships with men who will help raise those children. But reality is more complicated. We now know women cheat, too. Researchers have found that females of "monogamous" bird species seem to have affairs in order to land the best genes for their babies. Evolutionary scientists have long speculated that what applies to sparrows and roosters applies to human beings, too.

BREAKING UP

One night Rob didn't call Melissa after he said he would. It was unlike him, and she started to freak out with worry. Was he hurt? Was he with another woman? Melissa could feel her fear physically. Strangely enough, the state of romantic love can be reignited by the threat or fear of losing one's partner—of being dumped. Being dumped actually heightens the phenomenon of passionate love in the brain circuits of both men and women. That brain region desperately, hungrily seeks the loved one. Withdrawal—as if weaning from a drug—takes over. Moments of feeling as if your very survival is threatened occur, and a state of fearful alert is triggered in the amygdala. The anterior cingulate cortex—the part of the brain that engages in worry and critical judgment—starts to generate negative thoughts about losing the beloved. In this highly motivated, attentive state, obsessive thoughts of reunion take hold. This state elicits not trust and bonding, but painful, intense searching for the beloved. Melissa became crazed with thoughts of losing Rob. The part of herself that had become merged and expanded by his opinions, interest, beliefs, hobbies, mannerisms, and character was now in acute emotional, physical, and cognitive withdrawal, deep within the reward-driven areas of the brain.

The exhilarating expansion of the self that happened rapidly during the romantic-rush stage of love is now in a painful retraction. And when women experience betrayal or loss of love, they also respond differently than men do. When love is lost, abandoned men are three to four times more likely to commit suicide. Women, by contrast, sink into depression. Jilted females can't eat, sleep, work, or concentrate; cry all the time; withdraw from social activities; and *think* about suicide. My eighteen-year-old patient Louise, for example, had been inseparable for two years from her boyfriend, Jason, until the afternoon he left for college. He suddenly ended their relationship, telling her that he wanted to be free to date other girls while he was away. Four days later, I got an urgent call from Louise's father. She had been lying on the floor wailing inconsolably, not eating or sleeping, calling for Jason and moaning that she would rather die than to be without him.

Louise was hurting—literally—from the loss of love. Until recently, we thought that phrases like "hurt feelings" and "broken heart" were simply poetic. New brain-imaging studies, however, have revealed their accuracy. Rejection, it turns out, actually hurts like physical pain because it triggers the same circuits in the brain. Brain scans of people who have just been jilted by their beloveds also show the chemical shift from the high activity of romantic love to the flat biochemistry of loss and grief. Melissa wasn't quite to this point yet. Without love's surges of dopamine, the depression-despair response descends on the brain like a black cloud. This is what happened to Louise, but not to Melissa. Rob didn't even realize that he was supposed to call her that night and had gone out to play poker with the boys. When he realized how much he had hurt Melissa, he apologized and promised always to call her. This episode made both Melissa and Rob realize how essential they had become to each other and actually motivated them to take the next step toward making their relationship permanent. They got engaged.

It may be that the "brain pain" of lost love evolved as a physical alarm to alert us to the dangers of social separation.

Pain captures our attention, disrupts our behavior, and motivates us to ensure our safety and end our suffering. Given the importance for human survival of finding a mate, reproducing, and gaining food, nurturance, and protection, the pain of loss and rejection is likely hardwired in our brains so we'll avoid it—or at least move on quickly to another mate, who'll sweep us off our feet on a new, rapturous dopamine- and oxytocin-intoxicated high. What's the trigger for this high? Sex.

FOUR

Sex: The Brain Below the Belt

FINALLY, EVERYTHING WAS in place. Her mind was calm. The massage did the trick. Vacation was always the best place. No work, no worries, no phone, no e-mail. No place else for Marcie's brain to run. Her feet were even warm, and she wasn't thinking about getting up for a pair of socks. He was hot—and a great lover. She could let go and let it happen. Her brain's anxiety center was shutting down. The area for conscious decision making wasn't lighting up so intensely. The neurochemical and neurological constellations were aligning for orgasm. Blast off.

Female sexual turn-on begins, ironically, with a brain turnoff. The impulses can rush to the pleasure centers and trigger an orgasm only if the amygdala—the fear and anxiety center of the brain—has been deactivated. Before the amygdala has been turned off, any last-minute worry—about work, about the kids, about schedules, about getting dinner on the table—can interrupt the march toward orgasm.

The fact that a woman requires this extra neurological step may account for why it takes her on average three to ten times longer than the typical man to reach orgasm. So girls, tell your man to slow down and be patient, especially if you're trying to get pregnant. Research has shown that the biological reason for males coming more quickly is that females who orgasm after the male has ejaculated are more likely to conceive.

It's a delicate system, but the connection to the brain is about as direct as it gets. Nerves in the tip of the clitoris communicate straight to the sexual pleasure center of the female brain. When those nerves are stimulated, they boost electrochemical activity until it hits a threshold, triggers a burst of impulses, and releases bonding, feel-good neurochemicals such as dopamine, oxytocin, and endorphins. Ah, climax! If stimulation of the clitoris is cut off too soon, if the clitoral nerves aren't sensitive enough, or if fear, stress, or guilt interfere with stimulation, the clitoris is stopped dead in its tracks.

Marcie came to see me when she met John. She had had her first long, deep relationship, with Glenn, in her early twenties, but it didn't last, even though he was a good-looking guy and it had become a comfortable relationship in which she felt totally secure. She had really enjoyed their sex life and always had great orgasms with him, but he wasn't the man she wanted to marry. When she started dating again and hooked up with John, she found her body didn't respond as readily. It was not that John was a bad lover or had inadequate equipment. Just the opposite. He was more fun and even better looking than Glenn. But John wasn't Glenn, the man she'd grown comfortable and safe with. John was new, so she felt tense with him and couldn't have an orgasm. One day Marcie went to the doctor with a bad neck spasm, and he prescribed Valium to relax the muscle. She took a pill at dinner, and by the time she and John fell into bed and had sex, orgasm was no problem. The Valium had relaxed her brain, her amygdala was deactivated, and she was able to reach the neurochemical threshold of orgasm easily.

If you're not relaxed, comfortable, warm, and cozy, it's not likely to happen. In a brain-scan study of female orgasm, researchers discovered that the women needed to be comfortable and have their feet kept warm before they felt like engaging in sex. For many women, being relaxed—thanks to a hot bath, a foot rub, a vacation, or alcohol—improves their ability to have an orgasm, even with partners they don't feel completely comfortable with.

Women deeply in love and in the early phases of passion, who feel that their partners desire and worship them, are more likely to have easy orgasms. For some women, the state of security offered by a committed relationship or marriage can allow the brain to reach orgasm more easily than with a new person. As the orgasm subsides, waves of oxytocin cause a woman's chest and face to flush because the blood vessels expand. A glow of contentment and satisfaction surrounds her. Fear and stress are blocked out. But how this all happens remains a mystery to the men around us. Every woman has had the experience of lying in bed with a guy who asks, "Did you come?" Often, it's just hard for him to tell.

Because of the delicate psychological and physiological inter-connection, female orgasm has been elusive to confused male lovers—and to scientists. For decades women have volunteered to be prodded, filmed, tape-recorded, interviewed, measured, wired, and monitored by scientists. The shortened breath, arched back, warm feet, grimacing face, unintentional vocalizations, and jumping blood pressure of women's orgasm have all been measured. And now, because of MRI scans that show the activated and deactivated areas of the brain, we know much more about the female brain's control of orgasm.

If we took an MRI scan of Marcie's brain as she headed for the bedroom with John, we'd find that many of her brain circuits would be highly activated. As she snuggled down between the warm sheets, cuddled up to John, and started kissing and hugging, certain areas of her brain would become more calm and the areas for genital and breast sensitivity would begin to light up. As John began to touch her clitoris, her glowing brain areas would start to spark red, and as she grew more excited while he rubbed her clitoris, her brain area for worries and fear—the amygdala—would deactivate into a calm blue.

As she became more excited and pulled him inside her, the amygdala would completely deactivate and the pleasure centers would pulse red until—bingo—rapid, pulsing waves of orgasm flooded her brain and body.

For a man, orgasms are simpler. Blood has to rush to one crucial appendage for sexual climax to take place. For a woman, the neurochemical stars need to align. Most important, she has to trust who she's with.

Since the male model of arousal is basic hydraulics—blood flows to the penis, leading to erection—researchers have looked endlessly for the same simple mechanism in women. Doctors have surmised that women's arousal problems stem from low blood flow to the clitoris. There's never been any evidence, however, that this is true—and no researchers have ever found ways of measuring physical changes in the clitoris when it's aroused. Instead, they've groped for other indicators, such as lubrication, using clumsy methods such as weighing tampons before and after female research subjects watch erotic films. Scientific understanding of female sexual response is still decades—if not centuries—behind research on male erections, and the progress remains frustratingly slow. Even a recent anatomy textbook completely omitted a description of the clitoris while giving a three–page description of the penis. Medical doctors still feel that if a man can't get an erection, it is a medical emergency, but no one seems to feel the same urgency about sexual satisfaction for women.

Since Viagra's explosive debut in 1998, scientific interest in sex differences has heated up. Drug companies have been falling over themselves trying to find a pill or patch that can reliably kindle female desire. So far their efforts to discover a pink Viagra for women have been a bust. In 2004, Pfizer officially ended its eight-year quest to prove that Viagra boosted blood flow to the clitoris and therefore increased sexual enjoyment in women.

We now know for sure that, just as the female brain is not a smaller version of the male brain, the clitoris is not a little penis. The entire ring of tissues that surrounds the vaginal opening, the urethra, and the outer third of the vagina is connected by nerves and blood vessels to the tip of the clitoris—so all these tissues together are responsible for the excitation leading to orgasm. Some women refer to this area as their "ring of fire."

There is also no such thing as a vaginal versus a clitoral orgasm, as Freud erroneously thought. For nearly a century, his theory made women feel they were inadequate or not quite real women if they *only* had clitoral orgasms. Freud knew nothing at all, of course, about the anatomy of the clitoris or that of the female brain. Neuroscientists have discovered that the vagina is connected to the clitoris, and therefore the female orgasm is all from this one organ, which is connected to the pleasure centers in the brain. The clitoris really is the brain below the waist. The action, however, is not all below the waist, nor is it all guided by psychological factors. To the modern neuroscientist, the psychological and the physiological are not different—they are just opposite sides of the same coin.

IT DOESN'T TAKE MUCH TO SPOIL THE MOOD

Bad breath, too much slobber, a clumsy move with a knee, hand, or mouth, any little thing can spring the female amygdala back into action, cutting sexual interest and orgasm off at the pass.

Bad past experiences can start to occupy a woman's brain circuits, causing feelings of shame, awkwardness, or lack of safety. Twenty-eight-year-old Julie came to see me reporting she was unable to have an orgasm. She finally revealed that she had been molested by her uncle when she was a child and that the experience had made her dislike sex. She felt incredibly anxious when she had sex—even with her devoted, loving fiancé. Like Julie, four out of ten girls have had some kind of sexually upsetting experience in childhood that continues to occupy their brains with worries during adult sexual encounters—not being able to reach orgasm is one of the most common symptoms. Julie improved in her enjoyment of sex after getting both sex therapy and trauma therapy. Several months later she called me to report she'd had her first orgasm.

Especially for women, both biological and psychological factors influence arousability. Multitasking women end up having more distractions, which occupy their brain circuits and get in the way of sexual desire. Three months after she took a new job that required long hours, another patient of mine began having trouble reaching orgasm. She didn't have any downtime to relax with her husband, and she began faking orgasms to keep from hurting his ego. The worries and tension of her new job were interfering with her ability to relax, feel safe, and allow her amygdala to deactivate.

The interference of worry and stress with sexual satisfaction may also be one reason women like vibrators. A vibrator applied to the clitoris can often provide a faster, easier orgasm. You don't have to worry about the relationship, the guy's ego, whether he'll come too soon, or how you look in bed. Another patient of mine—divorced and in her forties—got so used to her vibrator that when she did become involved with a man again, she found he just wasn't doing as good a job as her mechanical device. Finally, she took drastic measures—she buried her vibrator in the backyard in order to force herself to get used to a real penis.

A woman needs to be put in the mood. Before sex, there has to be a soothing and smoothing of the relationship, and she has to be able to stop being annoyed with him. Anger at one's partner is one of the most common reasons for sexual problems. Many sex therapists say that, for women, foreplay is everything that happens in the twenty-four hours preceding penile insertion. For men, it's everything that happens three minutes before. Since many parts of a woman's brain are active at once, she must get into the mood by first relaxing and reconnecting positively with her partner. This is why she needs a good twenty-four hours to get in the mood, and why going on vacation is such an intense aphrodisiac. It allows her to unplug from daily life stress. So men, yes, bring out the flowers, chocolates, and sweet words—they work. A woman can't be angry at her man and want to have sex with him at the same time. And women, tell your men that if they plan on criticizing you or starting a fight on the day they are hoping to get lucky, they should think again.

They will have to wait for the twenty-four-hour clock to reset before you'll be ready.

THE FUNCTION OF FEMALE ORGASM

From an evolutionary perspective, male orgasm is no great mystery. It's little more than a biologically simple ejaculation accompanied by an almost addictive incentive to seek out further sexual encounters. The theory goes that the greater the number of inseminations a male achieves, the better are his chances of having his genes represented in future generations. Women's sexual climax is more complex and hidden—and can be easily faked. Women do not necessarily need to experience orgasm in order to conceive, though it helps.

Despite some scientists' belief that there is no purpose in female orgasm, it actually works to keep a woman lying down after sex, passively retaining sperm and increasing her probability of conception. Not to mention that orgasm is intense pleasure, and anything that feels good makes you want to do it again and again—just what Mother Nature had in mind. Others have suggested that female orgasm evolved to create a stronger partnership between lovers, inspiring in women feelings of intimacy and trust toward mates. An orgasm communicates a woman's sexual satisfaction with and devotion to a lover.

Many evolutionary psychologists have also come to view the female orgasm as a sophisticated adaptation that allows women to manipulate—even without their own awareness—which of their lovers will be allowed to fertilize her eggs. The quickened breath, moaning, racing heart, muscular contractions and spasms, and nearly hallucinatory states of pleasure that orgasm inspires may constitute a complex biological event with a functional design. Scientists believe orgasm may function as a "sperm competition," through which women's bodies and brains choose a winner.

The muscular contractions and uterine suction associated with women's orgasm have long been known to pull the sperm through the cervical mucus barrier. In one published account of the strength of the orgasmic suction into the cervix, a doctor reported that a patient's uterine and vaginal contractions during sex with a sailor had pulled off his condom. Upon inspection, the condom was found inside the tiny cervical canal. This means that the female orgasm can function to pull sperm closer to the egg. Scientists have discovered that when a woman climaxes any time between one minute before and forty-five minutes after her lover ejaculates, she retains significantly more sperm than if she does not have an orgasm. No orgasm means fewer sperm are sucked up into the cervix—the entry portal to the uterus, where the egg lies waiting. While a man worries about a woman's satisfaction with him as a lover—out of fear that she will stray or not want to have sex with him again—orgasmic females may actually be up to something far more clever. With her orgasms, a woman is deciding which partner will sire her children. If Marcie's Stone Age brain thinks John is sexy and good-looking enough to be a good genetic bet for her offspring, having an orgasm with him becomes serious business.

Biology has a way of winning out over our conscious minds by manipulating our reality to ensure evolutionary survival, so a woman's unconscious brain circuits will choose the best-looking guy, since he will give her bigger orgasms. Behavioral ecologists have also noted that female animals—from scorpion flies to barn swallows—prefer males with high degrees of bilateral body symmetry, which means that both sides of the body match. The reason perfectly matching body parts may be important is that the translation of genes into parts of the body can be perturbed by disease, malnutrition, or genetic defects. Bad genes or disease can cause deviation from bilateral symmetry in traits such as hands, eyes, and even birds' tail feathers, which are the visual features on which our female counterparts in the animal kingdom make their choices. Females want the best-looking guy to sire their offspring as well. The best males—those whose immune systems are strong, and who are healthy providers—develop with higher body symmetry. Females who choose symmetrical suitors are securing good genes for their offspring.

Humans share this preference. In studies, women consistently choose men whose faces, hands, shoulders, and other body parts are more symmetrical. This is not a matter of mere aesthetics. A large and growing body of medical literature documents that symmetrical people are physically and psychologically healthier than their less symmetrical counterparts. So if the guy you're dating is a little funny looking to you and you are put off, it may be nature signaling you about the quality of his genes. John just happened to be the best-looking man Marcie had ever dated, so maybe that had something to do with her desire to have him sire her children.

Scientists have reasoned that if women's orgasms are an adaptation for securing good genes for their offspring, then women should report more orgasms with good-looking, symmetrical mates. At the University of Albuquerque, researchers observed eighty-six sexually active heterosexual couples. Their average age was twenty-two, and the couples had been living together for two years—so trusting relationships had already been established. The researchers had each person privately—and anonymously—answer questions about his or her sexual experiences and orgasms. They then took photographs of each person's face and used a computer to analyze the features for symmetry. They also measured various body parts—the width of elbows, wrists, hands, ankles, feet, leg bones, and the length of the second and fifth fingers.

Indeed, the hypothesized relationship between male symmetry and female orgasm proved to be true. Reports provided by the women—and their lovers—indicated those whose partners were the most symmetrical enjoyed a significantly higher frequency of orgasms during sexual intercourse than those with less symmetrical mates.

Handsome men know this firsthand. Studies show that symmetrical men have the shortest courtships before having sexual intercourse with the women they date. They also invest the least time and money on their dates. And these handsome guys cheat on their mates more often than do guys with less well-balanced bodies. This is not what we women would like to believe. Instead, we like the bonding hypothesis, which says that women with kind, caring mates will have the most orgasms. But the reality is that men may just come in two different categories. There are the ones for hot sex and the ones for safety, comfort, and child rearing. Women are constantly longing for both wrapped into one package, but sadly, science shows that this may be wishful thinking.

Of course, no one is perfectly symmetrical, but we all do rate those with the greatest symmetry as being the best looking. To the researchers' surprise, women's romantic passion toward their mates did not increase the frequency of orgasm. Not only that but even though conventional wisdom holds that birth control and protection from disease increase orgasm rates among women—supposedly because they allow women to feel more relaxed during intercourse—no relationship emerged between female orgasm and the use of contraception. Instead, only how good looking the guy was correlated with a high frequency of female orgasm during copulation. After all, our brains are built for survival in the precontraceptive Stone Age. In evolutionary terms, condoms and the pill are just flashes in the pan—too recent to have changed the way we experience emotions or sex.

THE BIOLOGY OF FEMALE INFIDELITY

Mother Nature uses everything at her disposal to make sure couples get together and make babies, and this requires that sex happen at the right time of the month. Odors, for instance, are strongly linked to emotions, memory, and sexual behavior. Women's noses and brain circuits are particularly sensitive just before ovulation—not just to ordinary scents but also to the imperceptible effects of male pheromones. Pheromones are social chemicals that humans and other animals release into the air from their skin and sweat glands. They are found in male body sweat. Pheromones alter brain perceptions and emotions and influence desires—such as desire for sex. The brain changes its odor sensitivity as the estrogen surge leads up to ovulation. A small quantity of a pheromone is all it takes; the amount released in one one-hundredth of a drop of human sweat is enough to have a powerful effect. No wonder the perfume industry is going crazy trying to add this stuff to perfume and aftershave.

But what the scent industry doesn't know is that this effect depends on the day and even the hour of the menstrual cycle. When preovulatory women, for instance, who are at the peak of their monthly fertility, are exposed to a pheromone from male sweat glands called androstadienone (*an-dro-sta-DIE-en-own*, a close cousin of androstenedione, *an-dro-STEEN-die-own*, the major androgen made by the ovaries), within six minutes their mood brightens and their mental focus sharpens. These airborne pheromones keep women from getting into a bad mood for hours afterward. Beginning at puberty, only female brains, not male brains, are able to detect the androstadienone pheromone, and they're sensitive to it only during certain times of the month. It may be that androstadienone works on the emotions of females at their monthly reproductive peak to pave the way for social—and reproductive—interactions. It is interesting that Marcie mentioned to me during her first appointment that something about John's smell captivated her.

Using the body odor of men and the noses of women, Jan Havlicek of Charles University in Prague has hatched a controversial theory about pheromones and the female brain. He found that ovulating women who already have partners preferred the smell of other more dominant men but that single women showed no such preference. Havlicek argues that his findings support the theory that single women want nurturing men who will help raise a family. But once the home is secured, they have the biological urge to sneak around with men who have the best genes. Studies of mating patterns in species of birds once thought to partner for life showed that up to 30 percent of the baby birds were sired by males other than the ones taking care of them and living with their mothers.

Yet another blow to the myth of female fidelity is the dirty little secret in human genetic studies—up to 10 percent of the supposed fathers researchers have tested are not genetically related to the children these men feel certain they fathered. Ethical constraints prevent scientists from revealing this detail to anyone. Why does this happen? Is the female brain more likely to trigger an orgasm and conceive with a male who isn't her usual mate? Having an orgasm with a particularly desirable partner is thought to confer a reproductive advantage. Since a woman's orgasm sucks the sperm high up into the female reproductive tract, an orgasm with an enticing male gives a greater likelihood that the sperm will make it to the egg. This increased chance of conception with a sexy partner might be why women typically are more attracted to other men on the second week of their menstrual cycle—right before ovulation—their most fertile and flirtatious time of the month.

Another study found that women who have lovers on the side start to fake orgasm more often with their stable partners. Faking orgasm with their steady partners was more common even among women who reported only flirting with other men. Men are biologically geared to look for cues of sexual satisfaction for a reason—such satisfaction is reassurance about women's fidelity. Faking orgasm may function to distract a woman's primary partner from her infidelity. For men, faking sexual interest in their main partner is an old ruse to fool the women about the men's fidelity—sometimes over many years of a marriage. Researchers have shown that when women do engage in extramarital sex, they retain fewer sperm from their main partners (their husbands, in many cases) and experience more copulatory orgasms during their trysts, retaining more semen from their secret lovers. Taken together, these findings suggest that female orgasm is less about bonding with only the nice guys you want to marry than about a shrewd, subconscious, primitive evaluation of outside lovers' genetic endowment. Women are no more built for monogamy than men are. They are designed to keep their options open, and they fake orgasm to divert a partner's attention from their infidelities.

FUEL FOR LOVE

The sexual desire trigger for both genders is the androgen testosterone, the chemical that is mistakenly called, by some, the "male hormone." It's actually a sex and aggression hormone, and both men and women have lots of it. Men produce it in their testes and adrenal glands, while women make it in their ovaries and adrenal glands. In both males and females, testosterone is the chemical fuel that gets the brain's sexual engine going. When there's enough fuel, testosterone revs the hypothalamus, igniting erotic feelings and arousing sexual fantasies and physical sensations in the erogenous zones. The process works the same way in men and women, but there's a huge sex difference in the amount of testosterone that's available to "turn on" the brain. Men have on average ten to one hundred times more testosterone than women.

Even flirting is hardwired to testosterone. Studies have found that female rats with high testosterone levels are more playful than others and engage in more "darting" behavior, perhaps the rodent equivalent of sexual sassiness. In humans, the onset of sexual feelings and first intercourse for girls correlate with their testosterone levels. One study of eighth-, ninth-, and tenth-grade girls found that higher levels of testosterone were linked to more frequent sexual thoughts and more masturbation. Another study of adolescent girls revealed that rising testosterone was a significant predictor of first intercourse.

Despite the sharp rise in sexual interest for both teen girls and teen boys spurred by testosterone, there's still a significant difference in libido and sexual behavior. Between the ages of eight and fourteen, a girl's estrogen level increases ten to twenty times, but her testosterone level rises only about five times. A boy's testosterone level increases twenty-five-fold between ages nine and fifteen. With all that extra sexual rocket fuel, teen boys typically have three times more sex drive than girls of the same age—a difference that will persist through life. And while boys have a constantly rising level of testosterone through puberty, girls' sexual hormones ebb and flow each week— changing their sexual interest almost daily.

If a female's testosterone drops below a certain level, she'll lose sexual interest altogether. Jill, a forty-two-year-old premenopausal schoolteacher, had come to me complaining of no libido—which was causing marital problems. Her blood level of testosterone was very low, so I began treating her with testosterone therapy. To track her response to the hormone, I asked her to record how many sexual fantasies or dreams she had and how much she masturbated or was interested in masturbating. If we'd tracked only the number of times she had intercourse, that would most likely have given us a measure of her husband's libido. I asked her to come back in three weeks to assess her progress. During the time between appointments, Jill mistakenly doubled her dose of testosterone. Her face was blushing bright red when she came into the clinic. She sheepishly told me of her mistake and said her sexual urges were now so strong that she was running into the bathroom between classes to masturbate. She said, "This is becoming a real bother, but now I know what it must feel like to be a nineteen-year-old boy!"

If Jill had waited a little longer, another hormone in her menstrual cycle might have interfered with some of the flood of testosterone in her body. Testosterone is the main trigger the brain needs to ignite sexual desire, but it's not the only neurochemical that affects female sexual interest and response. Progesterone, which rises in the second half of the menstrual cycle, curbs sexual desire and acts to reverse partially the effect of testosterone in a woman's system. Some male sex offenders are even given injections of progesterone to decrease their sex drive. Women, too, have a decreased interest in sex when progesterone is high during the last two weeks of their menstrual cycle. Testosterone naturally rises—along with sexual urges—during the second week of the cycle, right before ovulation occurs at the peak of fertility. Estrogen does not cause increased sex drive by itself but peaks along with testosterone at the midpoint of the menstrual cycle. Estrogen tends to make females more receptive to sex and is essential for vaginal lubrication.

The Great Sexual Divide

The sex-related centers in the male brain are actually about two times larger than parallel structures in the female brain. When it comes to the brain, size does make a difference in the way women and men think about, respond to, and experience sex. Men, quite literally, have sex on their minds more than women do. They feel pressure in their gonads and prostates unless they ejaculate frequently. Males have double the brain space and processing power devoted to sex as females. Just as women have an eight-lane superhighway for processing emotion while men have a small country road, men have O'Hare Airport as a hub for processing thoughts about sex whereas women have the airfield nearby that lands small and private planes. That probably explains why 85 percent of twenty to thirty-year-old males think about sex many times each day and women think about it once a day—or up to three or four times on their most fertile days. This makes for interesting interactions between the sexes. Guys often have to talk women into having sex. It's not usually the first thing on women's minds.

These structural changes in the brain start as early as eight weeks after conception, when testosterone in the male fetus fertilizes the sex-related brain centers in the hypothalamus to grow larger. A second massive surge of testosterone at puberty then strengthens and enlarges other brain connections in the male that feed information to these sex centers, including the visual, smell, touch, and cognitive systems. The twenty-five-fold increase in testosterone between ages nine and fifteen fuels these larger sex connections in a male's brain for the rest of his youth.

Many of these structures and connections also exist in the female brain, but they're half the size. Females, from a biological point of view, simply devote less mental space to sexual pursuits. And their sexual interest ebbs and flows along with their monthly testosterone cycles. Male brain systems for sex are on alert with every waft of perfume and every female that walks.

What Women Don't Understand About What Sex Means to a Man

Jane and Evan, a thirty-something couple, came to see me with a familiar problem. Jane had just started a new job, gained some weight, and begun working extremely hard; she was putting all her time and energy—you might even say all her libido—into making a good impression at work. She found she simply wasn't in the mood to have sex anymore. Her husband was baffled, since when he had started his new, demanding job the year before, he had wanted sex even more than usual. Nonetheless, once Evan got Jane started, she enjoyed sex and could reach orgasm. She just never felt like getting started. It's the most common complaint among working women who come into my office.

It seems harmless enough: "Honey, I'm exhausted. I haven't eaten, work was really tough today, I'd love to cuddle in bed for a while, but really, I just want to eat, watch TV, and go to sleep. Is that okay?" He may say it is, but deep down, ancient wiring takes over. Remember, he's thinking about sex literally every minute. If she doesn't want to have sex, it can signal a waning of attraction or perhaps another man. In other words, the fading of love. Evan had insisted they come see me for some couple's counseling because he was convinced that Jane didn't love him anymore or, worse yet, that she was having an affair. As we discussed the differences between male and female brains, Jane realized that Evan's brain's reality had an unexpected reaction to her not wanting sex. His brain interpreted her lack of physical desire for him as "She doesn't love me anymore." Jane started being more sympathetic to what sex meant to her husband.

It's just like what happens with a woman and verbal communication. If her partner stops talking to her or responding emotionally, she thinks that he disapproves of her, that she's done something wrong, or that he doesn't love her anymore. She'll panic that she's losing him. She may even think he's having an affair. Jane truly was just tired and didn't feel attractive, but the thought took over in Evan's mind that she was falling out of love with him. He began to appear jealous and possessive as his biological reality made him search for the other man. If she wasn't having sex with him, she had to be doing it with someone else. After all, he would be. Once Jane understood all this, she told Evan what she had learned about sex being as important to a man as communication was to a woman, and she laughed when he said, "Great. Let's have more male communication."

Evan now understood that Jane needed more warm-up time, and Jane now understood Evan's need to be reassured that he was loved. And so they did have more "male communication." One thing led to another, and Jane became pregnant. Her reality was about to shift again, and sex—sorry, Evan—would move a little further down the list of things to do. The mommy brain was taking over.

FIVE

The Mommy Brain

MOTHERHOOD CHANGES YOU forever," my mother warned me. She was right. Long after my pregnancy, I'm *still* living and breathing for two—glued to my child, body and soul, by an attachment stronger than I ever thought possible. I'm a different woman since my child was born; and as a doctor, I appreciate why. Motherhood changes you because it literally alters a woman's *brain*—structurally, functionally, and in many ways, irreversibly.

It's nature's way, you could say, of ensuring the survival of the species. How else would you explain why someone like me— with absolutely no prior interest in children—felt born to be a mother after I came out of the drug-induced haze of a difficult labor? Neurologically, it was a fact. Deeply buried in my genetic code were triggers for basic mothering behavior that were primed by the hormones of pregnancy, activated by childbirth, and reinforced by close, physical contact with my child.

As in *Invasion of the Body Snatchers*—or, more accurately, *Invasion of the Brain Snatchers*—a mother is altered from within by the lovable little alien she bears. It's a trait we have in common with sheep, hamsters, monkeys, and baboons. Take, for example, a female Syrian hamster. Before she bears her young, she'll ignore or even eat helpless pups. As soon as she gives birth, she gathers up her wriggling newborns, keeps them fed and warm, and grooms and licks them to trigger body functions the pups need to ensure their own survival.

Humans are not quite so biologically determined. A woman's innate brain wiring, like that of other mammals, responds to basic cues—the growing of a fetus in her womb; her baby's birth; its suckling, touch, and smell; and frequent skin-to-skin closeness with her child. Even fathers, adoptive parents, and women who have never been pregnant can respond maternally after close, daily contact with an infant. These physical cues from the infant forge new neurochemical pathways in the brain that create and reinforce maternal brain circuits aided by chemical imprinting and huge increases of oxytocin. These changes result in a motivated, highly attentive, and aggressively protective brain that forces the new mother to alter her responses and priorities in life. She is relating to this person in a way she has never related to anyone else in her life. The stakes are life and death.

In modern society, where women are responsible for not only giving birth to children but working outside the home to support them economically, these changes in the brain create the most profound conflict of a mother's life. Nicole, a thirty-four-year-old investment banker, devoted years of her life to working hard in high school in order to get into Harvard University so that she could enter a prestigious career track giving her financial security and independence. Getting her Mrs. along with her B.A. was the furthest thing from her mind. After college she traveled the world, settled for a while into a job in San Francisco's financial district, then entered business school at the University of California, Berkeley. She spent four years there, getting a double master's in business administration and international relations in order to prepare herself for a career in the global economy. She finished Berkeley at twenty-eight and moved to New York, where she got a job as an associate at an investment bank.

The more you do something, the more cells the brain assigns to that task, and Nicole's circuits were becoming entirely focused on her job and career path. The next two years involved grueling but rewarding eighty-hour workweeks. She wanted to make her mark, and she put her mind, body, and soul into bonding with her career. But soon she met and fell in love with Charlie, a cute southern lawyer who worked across the hall from her; her brain began dividing the assignment of cells between her attachment to Charlie and her career. Thus, Nicole spent her early thirties learning to balance her relationship, which eventually turned into marriage, with her demanding job. There would soon be a third little person coming into her life, and the brain cells would be forced to divide again.

BABY ON THE BRAIN

Biology can hijack circuits in spite of our best intentions, and many women experience the first "mommy brain" symptoms long before they actually conceive a child, especially if they've been trying for a while. "Baby lust"—the deep-felt hunger to have a child—can hit a woman soon after she's cradled someone else's warm, soft newborn. Suddenly, even the least child-focused females can start craving the tender, delicious feel and smell of babies. They may chalk it up to ticking biological clocks, or the "me-too" influence of peers, but the real reason is that a brain change has occurred and a new reality has set in. The sweet smell of an infant's head carries pheromones that stimulate the female brain to produce the potent love potion oxytocin—creating a chemical reaction that induces baby lust. After visiting my sister's new baby, Jessica, for the first time when she was three months old, I was totally obsessed with babies for a long time. In a sense, I'd come down with an infectious condition that I'd caught—literally and physically—from my new niece: nature's sneak attack to trigger the desire to have a baby.

The mommy-brain transformation gets under way at conception and can take over even the most career-oriented woman's circuits, changing the way she thinks, feels, and what she finds important. Throughout pregnancy, a woman's brain is marinated in neurohormones manufactured by her fetus and placenta. Nicole was soon to experience firsthand the effects of these hormones. She and Charlie had barely come back from a weekend of love in upstate New York when it started to happen. If we had an MRI scanner looking inside Nicole's brain, we'd see just her normal female brain as the sperm penetrates the egg. Within two weeks after the egg has been fertilized, it implants firmly in the uterine lining and attaches to Nicole's blood supply. Once her blood supply and that of the fetus have been joined, hormonal changes begin in Nicole's body and brain.

Progesterone levels start to climb in Nicole's bloodstream and brain. She soon feels her breasts become tender and her brain getting sedated. We'd see her brain circuits become mellow as she got a sleepy feeling, which would make her need to rest and eat more than usual. Her brain's thirst and hunger centers are switched on full blast by the rising hormones. She will now need to produce up to double her normal blood volume. She never wants to be far from her water bottle, a faucet, or the bathroom. At the same time, her brain signals for eating, especially in the morning, become finicky as her brain is changing how it reacts to certain smells, especially of foods. She wouldn't want accidentally to eat something that would harm her fragile fetus during the first three months of pregnancy. This is why her brain is now overly sensitive to smell, which may make her nauseated most of the time. She may even get to the point of retching every morning—or at least feeling she would like to—all because her brain circuits for smell have changed massively thanks to the hormones of pregnancy.

Nicole struggles to get through every day during these first months of pregnancy. At work, all she can do is sit and stare at her stapler and try not to throw up. By the fourth month, however, a big transition occurs. Her brain has become accustomed to the massive hormone changes, and she can eat normally, even ravenously. Both her conscious and unconscious brain is now focused on what is going on in her uterus. As the fifth month rolls around, she starts to feel little gas bubbles in her abdomen; perhaps at first she thinks they are the usual gas gurgles from a big meal. But no, her brain is registering these as movements by her baby. The mommy brain has been hormonally primed for months, but not until now is Nicole consciously aware that she is growing a baby. She has been pregnant for almost half a year, and her brain has been changing and enlarging its smell circuits, thirst circuits, and hunger circuits, and putting the brakes on the pulsing cells in the hypothalamus that usually trigger her menstrual cycle. She is now ready for the love circuits to grow.

With each new kick or movement, she starts getting to know

her baby and longingly fantasizes about what it will be like to hold him or her in her arms. She can't quite imagine it but hungers for it nonetheless. This is also the first time Charlie may become interested in his growing child—feeling the kicks and listening to Nicole's abdomen for the little heartbeats. The baby may even tap-tap-tap back at him, and yes, fathers usually fantasize about a boy and mothers about a girl.

I remember those intense cravings for odd foods and feeling like I would surely vomit at even the whiff of greasy food. All these changes are brain signals that something or someone has invaded your system. Progesterone spikes from ten to a hundred times its normal level during the first two to four months of pregnancy, and the brain becomes marinated in this hormone, whose sedating effects are similar to those of the drug Valium.

This tranquilizing effect of progesterone and also high estrogen help protect against stress hormones during pregnancy. Those "fight-or-flight" chemicals, such as cortisol, are produced in large quantities by the fetus and placenta, so the mother's body and brain are flooded with them. By late pregnancy, the stress hormone levels in a woman's brain are as high as they would be during strenuous exercise. Oddly enough, though, these hormones don't result in the feeling of stress during pregnancy. Their impact is to make a pregnant woman vigilant about her safety, nutrition, and surroundings, and less attuned to other kinds of tasks, such as making conference calls and organizing her schedule. That's why, especially in the last month or so of pregnancy, Nicole starts to feel distracted, forgetful, and preoccupied. Not since puberty have there been so many changes going on in her brain at once. Of course, each woman's response depends on her psychological state and the events in her life, but these are the biological underpinnings of her changing reality during pregnancy.

At the same time, the size and structure of a woman's brain are changing, too. Between six months and the end of pregnancy, fMRI brain scans have shown that a pregnant woman's brain is actually shrinking. This may be because some parts of her brain get larger as others get smaller—a state that gradually returns to normal by six months after giving birth. In animal studies, we've seen that the thinking part of the brain, the cortex, enlarges during pregnancy, revealing the complexity and flexibility of female brains. Scientists still don't know exactly why brain size changes, but it seems to be an indication of the massive brain restructuring and metabolic changes going on. It's not that a woman is losing brain cells. Some scientists believe the mother's brain shrinks because of changes in cellular metabolism required for restructuring brain circuits—getting ready to turn some one-lane highways into superhighways. So while the body is gaining weight, the brain is actually losing it. In the final one to two weeks before giving birth, the brain begins to increase again in size as it constructs large networks of maternal circuits. Otherwise, the child's first sentence would have to be "Mommy, I shrunk your brain."

THE BIRTH OF THE MOMMY BRAIN

As her due date approaches, Nicole's brain will become almost exclusively preoccupied with her baby, and with fantasies about how she is going to make it through all the pain and physical effort to push out a healthy child without killing herself or the baby. Her mommy brain circuits switch to high alert. She gets bursts of energy even though she feels like a beached whale and can only waddle. Charlie also becomes preoccupied, not with the birthing process so much as with physical things—such as space for the baby, painting its room, and acquiring all the necessary equipment, most of which Charlie had purchased months ago. He suddenly thinks of six more things they will need. The daddy brain circuits are rapidly connecting up for the big event. Now the countdown to birth begins.

Nicole was given a due date but told it could be up to two weeks earlier or two weeks later. This is because each baby gets ready to be born at its own pace. This will be the first of many times Nicole and Charlie are held hostage to the innate timing of their child's developmental schedule, which rarely fits with what they have in mind.

The day finally arrives. Nicole's water breaks, and the amniotic fluid comes flooding down her legs. The baby is head down and ready. The mommy brain is switched on right at birth by a cascade of oxytocin. Cued by signals coming from the fully developed fetus when it is ready to be born, a pregnant woman's level of progesterone suddenly collapses, and oxytocin pulses flood her brain and body, causing the uterus to start contracting.

As the baby's head moves through the birth canal, more bursts of oxytocin fire in the brain, activating new receptors and forging thousands of new connections between neurons. The result at birth can be euphoria, induced by oxytocin and dopamine, as well as profoundly heightened senses of hearing, touch, sight, and smell. One minute you're sitting there, an awkward, beached whale, and the next your uterus is lurching into your throat and you can't believe that it's feasible to do the pelvic equivalent of expelling a watermelon through your nostril. After too many hours for most of us—the ordeal is over and your life and brain have changed forever.

In the mammal world, there's nothing unusual about these brain changes at birth. Take sheep, for instance. When a baby lamb passes through its mother's birth canal, oxytocin pulses rewire the ewe's brain in minutes, making it exquisitely sensitive to its baby's smell. For five minutes or less, just after birth, she's able to imprint the odor of her newborn. After that, she'll permit only her own lamb to nurse, rejecting others, which have unfamiliar smells. If she doesn't get to smell her own baby in those first five minutes, she will not recognize it and thus reject it, too. The act of birth triggers rapid neurological changes in the ewe that can be seen in her brain anatomy, neurochemistry, and behavior.

For the human mother, the lovely smells of her newborn's head, skin, poop, spit up breast milk, and other bodily fluids that have washed over her during the first few days will become chemically imprinted on her brain—and she will be able to pick out her own baby's smell above all others with about 90 percent accuracy. This goes for her baby's cry and body movements, too. The touch of her baby's skin, the look of its little fingers and toes, its short cries and gasps—all are now tattooed on her brain. Within hours to days, overwhelming protectiveness may seize her. Maternal aggression sets in. Her strength and resolve to care for and protect this little being completely grab the brain circuits. She feels as if she could stop a moving truck with her own body to protect her baby. Her brain has been changed, and along with it her reality. It is perhaps the biggest reality change of a woman's life.

Ellie, a thirty-nine-year-old first-time mother, had been happily married for two years to a self-employed salesman when she came to see me. During the first year of their marriage, she had lost a baby to miscarriage. Within six months she was pregnant again. Soon after her daughter's birth, she began having "freakouts," as she called them, about her husband's earning power and lack of health care benefits. In truth, their financial situation hadn't changed at all, and she'd never had any of these misgivings before. Now, however, she was blazingly angry with her husband for not providing a more secure home for her and their baby girl. Her needs and reality had changed radically, virtually overnight, and her new, protective maternal brain was tightly focused on her husband's ability to provide for the family.

With their aggressive, protective instinct fully primed, mothers become hypervigilant about all aspects of their home turf, especially infant safety, such as having babyproof electrical plug covers, installing latches on the kitchen cabinet doors, and making sure everyone washes their hands thoroughly before touching the baby. Like a human global positioning system, a mother's brain centers for sight, sound, and movement are honed in on monitoring and tracking her baby. This increased vigilance can take all forms, depending on the threat a mother sees to the safety and stability of her "nest." Even reassessment of her husband's role as provider is not unusual.

The maternal brain circuits change in other ways, too. Mothers may have better spatial memory than females who haven't given birth, and they may be more flexible, adaptive, and courageous. These are all skills and talents they will need to keep track of and protect their babies. Female rats, for example, that have had at least one litter are bolder, have less activity in the fear centers of their brains, do better on maze tests because they are better at remembering, and are up to five times more efficient in catching prey. These changes last a lifetime, researchers have found, and human mothers may share them.

This transformation holds true even for adoptive mothers. As long as you're in continuous physical contact with the child, your brain will release oxytocin and form the circuits needed to make and maintain the mommy brain.

THE DADDY BRAIN

Expectant dads go through hormonal and brain changes that roughly parallel those of their pregnant mates. This may explain the strange experience of my patient Joan. She and her husband, Jason, were ecstatic when their pregnancy test came back positive. Three weeks into the pregnancy, however, Joan began to have violent morning sickness. By her third month, she'd gradually improved—but then, to his surprise, Jason started feeling so nauseated in the morning that he couldn't eat breakfast and could barely drag himself out of bed. He lost five pounds in three weeks and was worried that he had a parasite. But what Jason actually had was Couvade Syndrome, a common complaint of expectant dads (up to 65 percent worldwide) who share some of the symptoms of pregnancy with their partners.

In the weeks before birth, researchers have found, fathers have a 20 percent rise in their level of prolactin, the nurturing and lactation hormone. At the same time, their level of the stress hormone cortisol doubles, increasing sensitivity and alertness. Then, in the first weeks after birth, men's testosterone plummets by a third, while their estrogen level climbs higher than usual. These hormone changes prime their brains for emotionally bonding with their helpless little offspring. Men with lower testosterone levels actually hear the cries of babies better. They don't hear quite as well as moms, however, when babies whimper, for example, fathers are slower than mothers to respond, although they tend to react just as quickly when a baby screams. Men's lower testosterone levels also decrease their sex drive during this time.

Testosterone suppresses maternal behavior, in females as well as males. Fathers who have Couvade Syndrome have higher levels of prolactin than other fathers do and steeper drops in testosterone when they interact with their babies. It may be, scientists think, that pheromones produced by a pregnant woman can cause these neurochemical changes in her mate, preparing him to be a doting father and equipping him—secretly, through smell—with some of the special nurturing mechanisms of the mommy brain.

HIJACKING THE PLEASURE CIRCUITS

Unlike sheep, most human females take longer than five minutes to bond with their newborns, but the window doesn't close that fast for humans. This is good news for women like me, who had less than ideal birth experiences, involving anesthesia, C-sections, or premature labor and delivery. By the time my son was born—after thirty-six hours of contractions, epidural anesthesia, and morphine—I was feeling a little dazed and only mildly curious to see the little guy. It wasn't the warm wave of gooey maternal love I had expected to feel immediately for my baby, partly because anesthesia and morphine mute the effects of oxytocin. Only after I emerged from my drugged state did I feel alert and protective. And I soon fell addictively, hopelessly in love with my new son, with all my maternal wiring and sensitivities fully firing.

"In love" is a phrase, in fact, that many mothers use to describe their feeling for their babies. And, not surprisingly, mother love looks a lot like romantic love on a brain scan. Researchers hooked new mothers up to brain-monitoring equipment and showed them photographs of their own children, then pictures of their romantic partners. The scans revealed that the same oxytocin-activated regions of the brain lit up in response to both photos. Now I know why I felt so passionately about my child, and why my husband was sometimes jealous. In both types of love, surges of dopamine and oxytocin in the brain create the bond, switching off judgmental thinking and negative emotions and switching on pleasure circuits that produce feelings of exhilaration and attachment. Scientists at University College in London found that the parts of the brain usually available for making negative, critical judgments of others—for example, the anterior cingulate cortex—are turned off when one is looking at a loved one. The tender nurturing response of the oxytocin circuits is reinforced by the feeling of pleasure created by bursts of dopamine, the pleasure and reward chemical. Dopamine is jacked up in the mommy brain by estrogen and oxytocin. This is the same reward circuit set off in a woman's brain by intimate communication and orgasm.

Being hopelessly in love with my baby soon became for me a permanent state of mind, reinforced daily. This is not to say the trials and tribulations of taking care of a new baby—such as having a whole day go by without having time for a shower and having gotten no sleep the night before—didn't get to me, too. (New mothers lose an average of seven hundred hours of sleep in the first year postpartum.) As Janet, one of my best girl-friends, who had just had a baby, too, commented, "Now you know why they say one kid and your life changes, two kids and your life is over." It's a good thing that in most cases the maternal pleasure button gets pushed over and over again, and the bonds grow tighter the longer the baby is close physically.

This increased bonding includes the effects of breast feeding. Most women who nurse their babies have an extra benefit: regular stimulation of some of the most pleasurable aspects of the mommy brain. In one study, mother rats were given the opportunity to press a bar and get a squirt of cocaine or press a bar and get a rat pup to suck their nipples. Which do you think they preferred? Those oxytocin squirts in the brain outscored a snort of cocaine every time. So you can imagine what a re-inforced behavior breast feeding is. It had to be good to guarantee the survival of our species. When a baby grasps its mother's breast with tiny hands and suckles on her nipple, it triggers explosive bursts of oxytocin, dopamine, and prolactin in the mother's brain. Breast milk then begins to flow. At first, all that tugging at your sore, bleeding nipples can make you think it will be impossible to get through another day of breast-feeding torture. But after a few weeks—if you haven't been driven to hara-kiri—you'll be able to quiet your screaming infant and calm yourself down by breast feeding. Within three or four weeks, the experience begins to become downright pleasurable. And not just because the pain has stopped. You start to look forward to breast feeding—unless you are so sleep-deprived that you can only go through your day in a dreamlike state. But at some point in the first few months, you may realize that breast feeding has become easy and you really, really enjoy it. Your blood pressure drops, you feel peaceful and relaxed, and you're basking in waves of oxytocin-inspired loving feelings for your baby.

Often mother love and breast feeding replace or interfere with a new mother's desire for her partner. Lisa came to see me a year after the birth of her second child. "Having sex," she told me matter-of-factly, "is no longer on my list of top ten things to do.

I'd much rather catch up on sleep or the million different chores I can never finish. But my husband is getting very irritable, even angry, that sex isn't a priority for me." When I asked Lisa how other things in her life were going, she described the wonderful feeling she got being physically close and skin to skin with her young children. Tears, in fact, welled up in her eyes when she told me how much she loves and feels "in love with" her youngsters. Her one-year-old was still breast feeding two or three times a day, and she said she could never have imagined such a complete, selfless sense of connection with another. "I love my husband," Lisa assured me, "but a lot of things are more important right now than taking care of his sexual needs. Sometimes I wish he'd just leave me alone."

Lisa's experience isn't unusual, and it's based on hardwired responses in her maternal brain. Lisa—like all women who are in skin-to-skin contact with babies and breast feeding—has a brain that's marinating in oxytocin and dopamine making her feel loved, deeply bonded, and physically and emotionally satisfied. It's no wonder that she has no need for sexual contact. Many of the positive feelings she usually gets from sexual inter-course are evoked, several times daily, by meeting the basic physical needs of her young children.

BREAST FEEDING AND THE FUZZY BRAIN

Every benefit has a cost, however, and one downside of breast feeding can be a lack of mental focus. Although a fuzzy-brained state is pretty common after giving birth, breast feeding can heighten and prolong this mellow, mildly unfocused state. Kathy, age thirty-two, came to see me frightened about the state of her memory. She was becoming increasingly absentminded and had even "forgotten" to pick up her seven-year-old son from school. She was still breast feeding her eight-month-old daughter and had noticed that she was getting more "ditsy" by the day. She told me, "What really worries me is that I'll go into a room to get something and forget what I was looking for—not once but up to twenty times a day." Kathy was particularly alarmed since her mother had Alzheimer's and she thought that these could be early symptoms of the illness. As we talked, Kathy remembered that she'd also been forgetful after the birth of her first child, and that the confusing state had passed soon after she she'd weaned her son.

The parts of the brain responsible for focus and concentration are preoccupied with protecting and tracking the newborn for these first six months. Remember, too, that besides lack of sleep, a woman's brain size returns to normal only at six months postpartum. Until then, as Kathy discovered, the level of mental fog can be alarming. A distinguished scientist I know was stunned, ten days after giving birth, to find that she couldn't even summon the basic words and phrases to hold an intelligent conversation. Several months later, however, once she had stopped breast feeding, she was as sharp as ever.

For most women, a little ditsiness may be a small price to pay for the benefits of nursing. And babies share the rewards. In fact, they're crucial partners in the neurological act of breast feeding. The hormones released by breast feeding and skin-to-skin contact spur the maternal brain wiring to forge new connections. The longer and more often a baby suckles, the more it triggers the prolactin-oxytocin response in the mommy brain. Pretty soon, a mother may feel her breasts tingling and leaking at the sight, sound, touch, or merely passing thought of nursing her baby. The immediate payoff for the infant is food and comfort. Oxytocin dilates blood vessels in the mother's chest, warming her nursing child, who also gets doses of feel-good compounds in the breast milk. The milk stretches the baby's stomach as it is fed releasing oxytocin in the baby's brain, too. This quiets and calms the baby—not just from the meal but from those relaxing waves of hormone.

Many mothers suffer "withdrawal" symptoms when they're physically separated from their babies, feeling fear, anxiety, and even waves of panic. It is now recognized that this is more than a psychological state but is a neurochemical state. I can remember returning to work when my son was five months old and packing my breast pump with me. The mommy brain, it turns out, is a finely tuned instrument, and separation, especially from a nursing baby, can upset a mother's mood, perhaps through a decline in the stress-regulating brain levels of oxytocin. I was a wreck on most days, but I thought it was just the stress of working at a full-time job at the hospital and trying to run a household, too.

Nursing mothers also go through withdrawal symptoms when they wean their babies. Since weaning often occurs in conjunction with returning to a stressful workplace, mothers may be flipped into an agitated, anxious state. Can you imagine how most breast-feeding mothers must feel at the end of eight hours or more at work? At home, they had oxytocin rushes flooding their brains every few hours from nursing their babies. At work, their usual supply is cut off, since oxytocin lasts only one to three hours in the bloodstream and brain. I can remember the intense desire I had by three o'clock most days to go home to my baby. Many mothers find they can ease these symptoms by pumping their breast milk at work for as long as possible. Then they can slowly taper off breast feeding and continue to nurse on evenings and weekends to maintain their breast milk supply. This allows them to still get the pleasurable oxytocin and dopamine boosts and to stay connected with their babies.

ONE GOOD MOMMY BRAIN DESERVES ANOTHER

The flip side of the warm, nurturing mommy experience is also common. In my practice it is not unusual to hear complaints about mothers. My thirty-two-year-old, newly pregnant patient Veronica immediately comes to mind. As she talked, it became apparent to me that her blistering anger toward her mother was directly linked to her busy mother's inattentive nurturing when Veronica was a child. Her mother would take off on business trips, leaving Veronica with a nanny for a week at a time, and whenever Veronica was upset, her mother seemed to shut down emotionally instead of offering warm support. She would say that she was too busy with work and tell Veronica to go play in the other room. Now that Veronica was pregnant with her first child, she was expressing fear that she might be the same kind of mother, given her busy, high-pressure job as an art director at a magazine. Two generations of working mothers unable to spend time with their children. Should she worry? Maybe.

Researchers have found that if mothers, for whatever reason—too many children, financial pressures, or careers that don't allow enough child-care time—cannot be good enough nurturers and are only weakly attached to their babies, they can negatively affect the trust and security circuits of their children. On top of that, females "inherit" their mothers' maternal behavior, good or bad, then pass it on to their daughters and granddaughters. Although behavior itself can't be passed on genetically, new research shows that nurturing capacity in mammals is passed on, in what scientists now call a nongenomic or "epigenetic"— meaning physically on top of the genes—type of inheritance. In Canada, the psychologist Michael Meaney discovered that a female rat born to an attentive mother but raised by an inattentive mother behaves *not* like her genetic mother but like the mother who raised her. The brains of the rat pups actually change according to the amount of nurturing they receive. Female pups showed the largest changes in brain circuits, such as the amygdala, that use estrogen and oxytocin. These changes directly affect the female rats' ability to nurture the next generation of pups. The mommy brain is built through architecture, not imitation. This inattentive mothering behavior can be passed on for three generations unless some beneficial change in the environment happens before puberty.

This finding has huge implications even if only some of it holds true for humans: how well you mother your daughter will determine how well she mothers your grandchildren. For many of us, the thought of being just like our mothers may be downright alarming, but already researchers are finding corresponding ties in humans between levels of mother-daughter bonding and the quality of care and strength of maternal bonds in the succeeding generation. Scientists also speculate that high levels of stress created between the demands of the workplace and the demands of the household can decrease the quality—not to mention the quantity—of nurturing care mothers are able to give their kids. And of course this behavior can affect not only children but grandchildren as well.

Scientists have also shown that high nurturing—from any loving, trust-inducing adult—may make babies smarter, healthier, and better able to deal with stress. These are qualities they will carry throughout their lives and into the lives of their own children. Children with less maternal care, by contrast, end up more easily stressed, hyperreactive, inattentive, sick, and fearful as adults. Studies of the brain effects of high-nurturing human mothers versus low-nurturing human mothers are few and far between, but one study showed that college-age adults who had low maternal care in childhood showed hyperactive brain responses to stress on PET scans. Researchers found these adults released more of the stress hormone cortisol into their bloodstreams than did their peers who had high maternal care in childhood. Those who had received low maternal care displayed increased anxiety, and their brains were more vigilant and fearful. This may be why Veronica always felt more easily stressed at work and during relationship challenges, and why she contemplated becoming a mother with such panic.

I often hear vivid stories about patients' grandmothers— about how they were able to be there for my patients who had an overwhelmed, busy, or depressed mothers. Veronica's paternal grandmother made her feel special, even though her maternal grandmother was as emotionally distant as her mother was. Veronica started to cry as she told me how her father's mother would drop preparations for a dinner party to color with her or to play dolls. Grandma made blueberry pancakes with warmed syrup and helped Veronica make her bed and clean her room. When there was a party to go to and Veronica needed clothes, this grandmother took her shopping and often let her buy dresses she loved but knew her mother would not have allowed.

If it happens often enough, this kind of special nurturing from any allomother—a substitute mom—can override the lack of nurturing from an overstressed mother. It's enough to break the cycle of nonattentive mothering, allowing the girl to provide more attentive nurturing to her own children. Veronica's paternal grandmother may have been the linchpin in creating generational change. Years later, when Veronica stopped by to introduce her new baby girl to me, it was clear that she had a loving bond with her daughter and had passed on not the negative example from her mother but the nurturing, trust-inducing one from her grandmother.

ATTENTION WORK DISORDER

Nicole, the Berkeley-MBA mom, was struggling with similar concerns when she came to see me. She had become so attached to her baby that she was having a meltdown about returning to work. She had a great job with terrific benefits, a high salary, and lots of opportunities for advancement, and she and her husband had incurred enough expenses that they needed the two incomes. She had to go back to work, and even though she had trouble imagining leaving her daughter in a stranger's hands, she unhappily did it.

Most mothers, on some level, feel torn between the pleasures, responsibilities, and pressures of children and their own need for financial or emotional resources. We know that the female brain responds to this conflict with increased stress, increased anxiety, and reduced brainpower for the mother's work and her children. This situation puts both kids and mothers in deep crisis every day. Nicole came back to see me just after her son turned three. She said, "My life's just not working anymore." She told me that her son was having bone-chilling, time-stopping tantrums in the grocery store when she had just two hours to figure out what to do with him and unpack the groceries before she slogged off to work. And that when he was sick and her husband was gone, she'd found herself praying at midnight that his fever would break by daylight so he could go to preschool and she would make it to her breakfast meeting—she'd been out a lot that winter with his illnesses and her boss's patience was wearing thin. There were also the endless stream of half days at school coming up, and she'd have to beg the nonworking mothers in her kid's class to take care of him until she got off work. She wasn't sure she or her son could take it anymore, but she couldn't afford to quit her job.

So is the working mother damned? Well, maybe yes and maybe no. In fact, one solution to these modern problems may come from our primate ancestors. As a rule, primates, including humans, are fairly practical about their investment in mothering. For example, primates in the wild are very rarely full-time mothers. Many mother monkeys balance infant care with their essential "work" of foraging, feeding activities, and resting. They also pitch in when needed to care for offspring other than their own—this is called alloparenting. In fact, in times of plenty, other moms easily adopt and care for foster children, even those from other communities or species. Many mammals have this capacity to bond with, nurture, and care for the offspring of others. An intriguing study of hunting among women of the Agta Negrito of Luzon (the Philippines) underscores the functions of networks of female kin. Women's hunting has largely been regarded as biologically impractical because hunting is assumed to be incompatible with the obligations of infant care. Specifically, hunting forays were thought to impair women's abilities to nurse, care for, and carry children. However, studies of cultures in which the females do hunt suggest exceptions that prove the rule. Agta women participate actively in hunting precisely because others are available to assume child care responsibilities. When women were observed to hunt, they either brought nursing children with them or gave the children to their mothers or oldest female siblings for care.

Mothering isn't necessarily a solo occupation by design in humans—or restricted to the birth mother in an urban environment either. From the child's perspective, nurturing is nurturing, no matter which loving, security-inducing caregiver it comes from. Nicole was able to negotiate a more flexible schedule at work so that her son could attend preschool half days along with his friend who lived next door and the two mothers covered for each other.

IDEAL MOMMY BRAIN ENVIRONMENT

One environmental factor that is essential for good mothering in any animal is predictability. It's not about how many resources are available, it's about how regularly they can be obtained. In one study, mother rhesus monkeys were set up with their youngsters in three different environments: one had plenty of food every day, one had scarce food every day, and the third had plenty of food on some days but scarce food on others. The amount of nurturing behavior mothers gave to their youngsters in these environments was recorded every hour on video. Youngsters in the best environment, with plenty of food, got the most responsive nurturing from their moms, while those in the environments with scarce but steady amounts of food got almost as much. But those from the unpredictable environment not only got the least amount of nurturing but received abusive and aggressive attacks from their moms. The mother and infant monkeys in the unpredictable environment had higher levels of stress hormones and lower levels of oxytocin than their peers in the other environments.

In an unpredictable human environment, mothers become fearful and timid, and babies show signs of depression. The youngsters cling to their moms and are much less interested in exploring and playing with others—traits that linger on into adolescence and adulthood. This study supports the common-sense notion that mothers can do their best in a predictable environment. According to the primatologist Sarah Hrdy, humans evolved as cooperative breeders in settings where mothers have always relied on allomaternal care from others. So whatever a mother does and others do to help her, inside or out-side the home, to ensure the predictability and availability of resources—financial, emotional, and social—may ultimately secure her children's future well-being.

LIVING FOR TWO

I can remember how stunned I was to discover that my inde-pendent and self-sufficient lifestyle no longer worked after I had a child. My thinking had always been that I could organize it myself and do most of the mothering alone. Boy was I wrong. Since a mother's brain has virtually expanded its definition of the self to include her child, the needs of the child will become a biological imperative for the mother, perhaps more compelling to her brain than her own needs. I could no longer schedule my life so neatly. I didn't know how much help from others, besides my husband's, I would need. Every new mother needs to under-stand the biological changes that are going to happen in her brain and then plan out her pregnancy and mothering dynamic in advance. This life challenge can stimulate your brain circuits to grow like no other. Setting up a predictable environment for work and for loving, security-inducing child care will be crucial. A mother's emotional and mental development depends to a great extent on the context in which she mothers. Knowing that you will need extra support for yourself and some good allomothers for your child will be key to your success as a mother. If we can provide a reliable, secure environment for the mommy brain, we can stop the domino effect of stressed mothers and insecure, stressed children.

The changes that happen in the mommy brain are the most profound and permanent of a woman's life. For as long as her child is living under her roof, her GPS system of brain circuits will be dedicated to tracking that beloved child. Long after the grown baby leaves the nest, the tracking device continues to work. Perhaps this is why so many mothers experience intense grief and panic when they lose day-to-day contact with the person their brain tells them is an extension of their own reality.

Developmental psychologists believe that the female brain's extreme ability to connect through reading faces, interpreting tones of voice, and registering the nuances of emotion are traits that were selected evolutionarily from the time of the Stone Age. These traits make it possible for the female brain to pick up cues from nonverbal infants and anticipate their needs. The female brain will turn this extraordinary ability on all her relationships. If she's married or partnered with a male brain, each will inhabit two different emotional realities. The more both know about the differences in the emotional realities of the male and female brains, the more hope we have of turning those partnerships into satisfying and supportive relationships and families. Just what the mommy brain needs to be at its best.

SIX

Emotion: The Feeling Brain

IS THERE ANY TRUTH to the cultural stereotype that women are more emotionally sensitive than men? Or that a man wouldn't know an emotion unless it hit him on the head? My husband said we didn't need a separate chapter on emotions. I didn't see how I could write this book without one. The explanation of our different mind-sets lies in the biology of our brains.

My patient Sarah was positive that her husband, Nick, was seeing another woman. Over several days she silently chewed on the idea. First, she felt unsure of what she suspected. Then, as her mind worked over her anger at the possibility that he was cheating, her *gut sense* of betrayal became overwhelming. She stopped smiling. How could he do this to her and their baby daughter? She moped around the house. She couldn't understand why her husband never tried to cheer her up. Couldn't he see how miserable she was?

Nick had always been larger than life to her—so talented and smart—Sarah felt honored to be his wife. When he turned his beam of brilliance on her to tell her his deepest thoughts, she felt she was eliciting greatness from him. She lived for the moments he shined on her. But when it came to an emotional interaction, it was a different story. He was a little hard to reach. So one night, when she burst into tears over dinner, Nick was stunned. Sarah couldn't figure out why he was so surprised. She'd been showing a cold face to him for days. She went back over all of those moments when he shined on her with such intensity and how wonderful this always made her feel—that he really loved her and cared about her. Was she wrong about that—or didn't she please him anymore? How could he be so insensitive to her emotional state?

Imagine for a moment that we had an MRI scanner. This is what it might look like inside Sarah's brain and body as she processed her conversation with Nick: As she asks him if he is seeing someone else, her visual system begins scanning Nick's face intently for signs of his emotional response to her question. Does he tighten his face or relax it? Does he clench around the mouth, or keep it neutral? Whatever the expression on his face, her eyes and facial muscles will automatically mimic it. The rate and depth of her breathing start to match his. Her posture and muscle tension conform to his. Her body and brain receive his emotional signals. This information is sent through her brain circuits to search her emotional memory banks for a match. This process is called "mirroring," and not all people can do it equally well. Although most of the studies on this topic have been done on primates, scientists speculate that there may be more mirror neurons in the human female brain than in the human male brain.

Sarah's brain will begin stimulating its own circuits as if her husband's body sensations and emotions were hers. In this way, she can identify and anticipate what he is feeling—often before he is conscious of it himself. Matching breathing, matching posture, she is becoming a human emotion detector. She is feeling his tension in her gut, his jaw clenching in the strain of her neck. Her brain registers the emotional match: anxiety, fear, and controlled panic. As he starts to speak, her brain carefully searches to see if what he says is congruent with his tone of voice. If the tone and meaning do not match, her brain will activate wildly. Her cortex, the place for analytical thinking, would try to make sense of this mismatch. She detects a subtle incongruence in his tone of voice—it is a little too over-the-top for his protestations of innocence and devotion. His eyes are darting a bit too much for her to believe what he is saying. The meaning of his words, the tone of his voice, and the expression in his eyes do not match. She knows: he is lying. She is now using her brain's entire emotion network as well as her cognitive and emotional suppression circuits to keep from crying. But the dam breaks. Tears roll down her cheeks. Nick's face looks puzzled. He has not been following Sarah's emotional nuances—otherwise he would have known she was losing it.

Sarah was right. When Nick came to see me as part of couples counseling, he revealed that he had been spending lots of time with a female co-worker. The relationship hadn't been consummated, but he had crossed the line in his flirtations and was becoming emotionally involved. Sarah knew it, literally, in every cell in her body, but since he hadn't technically cheated, Nick figured he was in the clear. When he realized that Sarah had correctly identified what he was feeling and thinking, he once again thought he was married to a psychic, but she was just doing what the female brain is expert at: reading faces, interpreting tone of voice, and assessing emotional nuance.

Maneuvering like an F-15, Sarah's female brain is a high-performance emotion machine—geared to tracking, moment by moment, the non-verbal signals of the innermost feelings of others. By contrast, Nick, like most males, according to scientists, is not as adept at reading facial expressions and emotional nuance—especially signs of despair and distress. It's only when men actually see tears that they realize, viscerally, that something's wrong. Perhaps that's why women evolved to cry four times more easily than men—displaying an unmistakable sign of sadness and suffering that men can't ignore. Couples like Nick and Sarah come to see me all the time for counseling. She complains about his lack of emotional sensitivity—because hers is so finely tuned—and he complains about the fact that she doesn't seem to realize he loves her. These are the different realities of the male and female brains at work.

THE BIOLOGY OF GUT FEELINGS

Women know things about the people around them—they feel a teenage child's distress, a husband's flickering thoughts about his career, a friend's happiness in achieving a goal, or a spouse's infidelity at a gut level.

Gut feelings are not just freefloating emotional states but actual physical sensations that convey meaning to certain areas in the brain. Some of this increased gut feeling may have to do with the number of cells available in a woman's brain to track body sensations. After puberty, they increase. The estrogen increase means that girls feel gut sensations and physical pain more than boys do. Some scientists speculate that this greater body sensation in women punches up the brain's ability to track and feel painful emotions, too, as they register in the body. The areas of the brain that track gut feelings are larger and more sensitive in the female brain, according to brain scan studies. Therefore, the relationship between a woman's gut feelings and her intuitive hunches is grounded in biology.

When a woman begins receiving emotional data through butterflies in her stomach or a clench in the gut—as Sarah did when she finally asked Nick if he was seeing someone else—her body sends a message back to the insula and anterior cingulate cortex. The insula is an area in an old part of the brain where gut feelings are first processed. The anterior cingulate cortex, which is larger and more easily activated in females, is a critical area for anticipating, judging, controlling, and integrating negative emotions. A woman's pulse rate jumps, a knot forms in her stomach—and the brain interprets it as an intense emotion.

Being able to guess what another person is thinking or feeling is, essentially, mind reading. And overall, the female brain is gifted at quickly assessing the thoughts, beliefs, and intentions of others, based on the smallest hints. One morning at breakfast, my patient Jane looked up to see that her husband, Evan, was smiling. He held the newspaper, but his gaze was lifted and his eyes darted back and forth, though he wasn't looking at her. She had seen this behavior many times before in her lawyer husband and asked, "What are you thinking about? Who are you beating in court right now?" Evan responded, "I'm not thinking about anything." But in fact he was unconsciously rehearsing an exchange with counsel he might be having later that day—he had a great argument and was looking forward to mopping up the courtroom with his opponent. Jane knew it before he did.

Jane's observations were so minute that to Evan she appeared to be reading his mind. This often unnerved him. Jane had watched Evan's eyes and facial expression and correctly inferred what was going on in his brain. And later, when he seemed to display hesitancy—a slight pause before speaking, tightness in his mouth, a low and flat tone of voice—when talking about going to the office, she sensed that a big career shift was coming. She mentioned this, but Evan said he hadn't been thinking about anything like that. A few days later, he announced he wanted to leave his firm and become a judge. Jane's observations were being made subconsciously, so these thoughts didn't register as anything but gut feelings.

Men don't seem to have the same innate ability to read faces and tone of voice for emotional nuance. This difference was in abundant display during the first few weeks after Jane and Evan met. She told me he was going way too fast for her, but he was unaware of her discomfort. A female friend of Evan's took one look at Jane, spotted her uneasiness, and warned Evan to back off. He didn't listen, and the results were nearly disastrous.

In that moment, Evan's female friend established emotional congruence with Jane, something that women seem to do naturally and that has been found to be crucial for successful psychotherapy. A study at California State University, Sacramento, of psychotherapists' success with their clients showed that therapists who got the best results had the most emotional congruence with their patients at meaningful junctures in the therapy. These mirroring behaviors showed up simultaneously as the therapists comfortably settled into the climate of the clients' worlds by establishing good rapport. All of the therapists who showed these responses happened to be women. Girls are years ahead of boys in their ability to judge how they might avoid hurting someone else's feelings or how a character in a story might be feeling. This ability might be the result of the mirror neurons firing away, allowing girls not only to observe but also to imitate or mirror the hand gestures, body postures, breathing rates, gazes, and facial expressions of other people as a way of intuiting what they are feeling.

The cat is out of the bag now. This is the secret of intuition, the bottom line of a woman's ability to mind-read. Nothing mystical at all. In fact, brain-imaging studies show that the mere act of observing or imagining another person in a particular emotional state can automatically activate similar brain patterns in the observer—and females are especially good at this kind of emotional mirroring. Through this kind of approximation, Jane figured out how Evan felt because she could feel through her body sensations.

Sometimes, other people's feelings can overwhelm a woman. My patient Roxy, for example, gasped every time she saw a loved one hurt him or herself—even when they did something as minor as stub a toe—as if she were feeling their pain. Her mirror neurons were overreacting, but she was demonstrating an extreme form of what the female brain does naturally from childhood and even more in adulthood—experience the pain of another person. At the Institute of Neurology at University College, London, researchers placed women in an MRI machine while they delivered brief electric shocks, some weak and some strong, to their hands. Next, the hands of the women's romantic partners were hooked up for the same treatment. The women were signaled as to whether the electric shock to their beloveds' hands were weak or strong. The female subjects couldn't see their lovers' faces or bodies, but even so, the same pain areas of their brains that had activated when they themselves were shocked lit up when they learned their partners were being strongly shocked. The women were feeling their partners' pain.

Like walking in another's brain, not just his shoes. Researchers have been unable to elicit similar brain responses from men.

Many evolutionary psychologists have speculated that this ability to feel another's pain and quickly read emotional nuance gave Stone Age women a heads-up to sense potential dangerous or aggressive behavior and thus avoid the consequences to themselves and protect their children. This talent also primes women for anticipating the physical needs of nonverbal infants.

Being this emotionally sensitive has its pros and cons. Jane, a normally brash and courageous person, told me that she could not get to sleep for hours after seeing an intense action flick. In a study on the aftereffects of frightening films, women were more likely to lose sleep than men. Studies show that, from childhood, females startle more easily and react more fearfully as measured through electrical conductivity in the skin. Evan had to readjust his movie-watching habits if he wanted to include Jane. So when he suggested that they watch *The Godfather*, he made sure it was in the middle of the day.

GETTING THROUGH TO THE MALE BRAIN

In the male brain, most emotions trigger less gut sensation and more rational thought. The typical male brain reaction to an emotion is to avoid it at all costs. To get a male brain's emotional attention, a woman needs to do the equivalent of yelling, "Periscope up! Emotion coming. All hands on deck!"

It took a lot for Jane to get the message to Evan that he was moving too fast when they met. Jane explained to me that she had been burned in relationships before and was seriously gun-shy when she started dating Evan. He paid no attention to the signals she was sending that she was a bona fide commitment phobe. On the third date, he told her he thought she was the one. By the second week, he wanted them to move in together and plan for the future. When Jane came in for her session that week, she looked as scared as a deer caught in headlights. Then, over pizza during the third week, Evan let her know he wanted to get married and start a family, and he was sure she was the one he wanted to do it with. Jane promptly turned green and ran to the bathroom. It wasn't until she showed obvious signs of distress that Evan realized he was moving too fast. He hadn't heeded the earlier warning from his female friend and now he was in deep trouble.

Bursting into tears often grabs the male brain's attention, but the tears nearly always come as a complete surprise—and extreme discomfort—to a man. A woman, because of her expert ability to read faces, will recognize the pursed lips, the squeezing around the eyes, and the quivering corners of the mouth as preludes to crying. A man will not have seen this buildup, so his response is usually "Why are you crying? Please don't make such a big deal out of nothing. Being upset is a waste of time." Researchers conclude that this typical scenario means the male brain must go through a longer process to interpret emotional meaning. Most men just don't want to take the time to figure out the emotion, and they become impatient because it takes longer for them. Simon Baron-Cohen at the University of Cambridge believes this is what happens in men with the extreme male brain that is characteristic of Asperger's disorder. These men become unable to look at a face, let alone read it. The amount of emotional input from another person's face registers on their brains as unbearable pain.

Tears in a woman may evoke brain pain in men. The male brain registers helplessness in the face of pain, and such a moment can be extremely difficult for them to tolerate. The first time Jane cried in front of an otherwise very affectionate Evan, she was stunned that she got a perfunctory hug and a few pats on the back followed by "Okay, that's enough." This seemingly rejecting behavior became a bone of contention in their relationship. The two came to see me for an urgent couples session. Evan needed to communicate to Jane that seeing her cry was nearly impossible for him to bear because when he saw her in pain he felt powerless to do anything about it. Slowly, they began to work on a compromise, so that Jane could get the comfort she needed and Evan could ease the pain he experienced. When Jane was upset, Evan would sit on the couch with a box of Kleenex on his lap. He would cradle her with one arm and hold a magazine or book with the other in order to distract himself from his own discomfort. After a few years, Evan was able to recognize when Jane needed a good cry, and soon he could simply hold her and take care of her until she was done.

WHEN HE DOESN'T RESPOND THE WAY SHE WANTS HIM TO

Being able to "be there" during emotionally difficult times is hardwired into women, which is why they are often baffled by their husbands' inability to sit with sadness or despair. One study showed that newborn girls, less than twenty-four hours old, respond more to the cries of another baby—and to human faces—than do boys. Girls as young as a year old are more responsive to the distress of other people, especially those who look sad or are hurt. Men pick up the subtle signs of sadness in a female face only 40 percent of the time, whereas women can pick up these signs 90 percent of the time. And while men and women are both comfortable being physically close to a happy person, only women report that they feel equally comfortable being close to someone sad.

Think of your girlfriends who will stick with you when you are hurt or sad. They'll ask you when it happened, what was said, if you've been able to sleep or eat, and "do you need me to come over?" To them, the details are important. I remember when I broke my ankle a few years ago and my girlfriends would just stop by and bring me some little treat they knew I'd like. They did everything they could to keep me from getting cabin fever. They knew how to help. Guy friends, by contrast, offered a quick "I hope you feel better" before jumping off the phone or walking out the door. It's not that they were being insensitive on purpose. It may be more about ancient wiring. Men are used to avoiding contact with others when they themselves are going through an emotionally rough time. They process their troubles alone and think women would want to do the same. Periscope down; submarine dives twenty fathoms to solve it alone.

The same apparent insensitivity can show up during other emotional exchanges. Jane and Evan moved in together, and after a few pressure-free months, Jane realized she wanted to spend the rest of her life with Evan, too. She decided to let him know. After two months of her dropping hints—about kids, about buying a house together, about what city they'd finally settle in—Evan didn't do anything. At our next session, Jane reported to me that, panicked, she went for the direct route: "I'm ready to get married," she told him one afternoon. Evan replied, "Okay, that's good to know," then went to watch the basketball playoffs. Jane began to panic. Had he changed his mind? Did he not love her anymore? She chased him around the house for three hours, haranguing him. Out of utter frustration and humiliation, she burst into tears, asking him if he was thinking of leaving her. "What?" Evan exclaimed. "How did you come to that conclusion? This is the first time you've given me any indication that you're ready. I was going to buy a ring and make a nice romantic dinner plan, but I can see you're not going to let me do that. So okay. Will you marry me?" Jane couldn't understand how he had missed the signs that she was ready, and Evan couldn't understand why she was so upset that he didn't answer right away.

Remember the little girl who wouldn't rest until she got an expression out of a mime? If she doesn't get the expected response, she will persist until she begins to conclude that she's done something wrong or that the person doesn't like or love her anymore. Something similar was playing out for Jane. When Evan didn't immediately ask her to marry him, and didn't respond to her direct approach, she concluded that he didn't love her anymore. Evan, in fact, was just trying to buy time to do things right.

EMOTIONAL MEMORY

It would be interesting to follow Evan and Jane over the years and see how they remember these early days. Most likely, his version, through no fault of his own, will be the movie trailer. Hers will be the full-length motion picture. She will take this as a sign of his waning love. When she expresses this reaction to him, he won't know what she's talking about. To understand their differences, we have to look at how emotions get stored as memories in the female brain.

Picture, for a moment, a map showing the areas for emotion in the brains of the two sexes. In the man's brain, the connecting routes between areas would be country roads; in the woman's brain, they'd be superhighways. According to researchers at the University of Michigan, women use both sides of the brain to respond to emotional experiences, while men use just one side. They found the connections between the emotion centers are also more active and extensive in women. In another study, at Stanford University, volunteers viewed emotional images while having their brains scanned. Nine different brain areas lit up in women, but only two lit up in men. Research also shows that women typically remember emotional events—such as first dates, vacations, and big arguments—more vividly and retain them longer than men. Women will know what he said, what they both ate, if it was cold outside or it rained on their anniversary, while men may forget everything except whether or not she looked sexy.

For both sexes, the emotional gatekeeper is the amygdala, an almond-shaped structure located deep within the brain. The amygdala is like the brain's Homeland Security Alerting and Coordinating System, switching on the rest of the body systems—the gut, skin, heart, muscles, eyes, face, ears, and adrenal glands—to look out for incoming emotional stimuli. The first relay station for emotion from the amygdala to the body is the hypothalamus. Like the Joint Chiefs, it's responsible for co-ordinating the launch of systems that raise blood pressure, heart rate, and breathing rate, and stimulate the fight-or-flight reaction after receiving reports from the body. The amygdala also alerts the cortex, the brain's Intelligence Branch, which sizes up the emotional situation, analyzes it, and determines how much attention it deserves. If it senses enough emotional intensity, the cortex cues the amygdala to alert the conscious brain to pay attention. This is the moment when we're flooded with conscious emotional feeling. Before this point, all this brain processing is happening behind the scenes. The brain's decision-making center, or Executive Branch—the prefrontal cortex—can now decide how to respond.

Part of the reason that her memory is better for emotional details is that a woman's amygdala is more easily activated by emotional nuance. The stronger the amygdala response to a stressful situation, such as an accident or threat, or a pleasant event, such as a romantic dinner, the more details the hippocampus will tag for memory storage about the experience. Scientists believe that because women have a relatively larger hippocampus, they have better memories for the details of both pleasant and unpleasant emotional experiences—when they happened, who was there, what the weather was like, how the restaurant smelled—in a detailed, three-dimensional, sensory snapshot.

Thirteen years later, Jane remembers every minute of the day she and Evan decided to get married, but as time wore on, Evan began to forget how it happened. They used to laugh about it all the time, but now he looks at her blankly when she recounts the details. He remembers that she got sick the first time he mentioned marriage, but he doesn't remember how he eventually asked her. He didn't store in memory any of these precious details. This is not because Evan doesn't love Jane; it's because his brain circuits are incapable of retaining the information, so it doesn't encode in his long-term memory. If she had activated his amygdala with a threat to the relationship or a physical danger, the memory would have been burned into his circuits just as it is into hers.

There are two exceptions in which men register emotions and thus detailed memories. If the person he is interacting with is blatantly angry and threatening, a man will be able to read that emotion as quickly as a woman can. His response to an aggressive threat will be as quick as hers, and will trigger an almost instant muscular reaction. Threatening to leave or threatening him physically will get his attention in an instant. Jane told me that, though she didn't mean it, she had told Evan during an argument that she couldn't take his stubbornness anymore and she was leaving. Evan was so traumatized that he asked her never to threaten to leave unless she really meant it. That was an argument he never forgot.

THE FEMALE BRAIN'S TOUGH TIME WITH ANGER

Another major difference between the male and female brains is in how they process anger. Although men and women report feeling the same amount of anger, the expression of anger and aggression is clearly greater in men. The amygdala is the brain center for fear, anger, and aggression, and it's physically larger in men than in women, whereas the anger, fear, and aggression control center—the prefrontal cortex—is relatively larger in women. As a result, it's easier to push a man's anger button. The male amygdala also has many testosterone receptors, which stimulate and heighten its response to anger, especially after the testosterone surges at puberty. That's why men whose testosterone levels are high, which includes younger men, have short anger fuses. Many women who start taking testosterone also notice that their anger response is suddenly quicker. As men age, their testosterone naturally declines, the amygdala becomes less responsive, the prefrontal cortex gains more control, and they don't get angry as fast.

Women have a much less direct relationship to anger. I grew

up hearing from my mother that the quality and longevity of a marriage could be measured by the number of bite marks on a woman's tongue. When a woman "bites her tongue" to avoid expressing anger, it's not all socialization. A lot of it is brain circuitry. Even if a woman wanted to express her anger right away, often her brain circuits would attempt to hijack this response, to reflect on it first out of fear and anticipation of retaliation. Also, the female brain has a tremendous aversion to conflict, which is set up by fear of angering the other person and losing the relationship. This may be accompanied by a sudden change in some brain neurochemicals, such as serotonin, dopamine, and norepinephrine—causing an unbearable activation in the brain on almost the same spectrum as a seizure—when anger or feelings of conflict arise in a relationship.

Perhaps in response to this extreme discomfort, the female brain developed an additional step in processing and avoiding conflict and anger, a series of circuits that hijack the emotion and chew on it, the same way a cow has an extra stomach that rechews its food before it is digested. These extralarge areas in the female brain are the prefrontal cortex, and the anterior cingulate cortex. They are the female brain's version of the extra stomach for chewing on anger. As we saw earlier, women activate these areas more than men at the fear of loss or pain. In the wild, the loss of a relationship with a protective male provider could have spelled doom. Cautiously holding her anger back may also have saved a female and her offspring from retaliation from men—if she didn't fly off the handle, she was less likely to evoke an extreme response from a trigger-tempered male.

Studies show that when a conflict or argument breaks out in a game, girls typically decide to stop playing to avoid any angry exchange, while boys generally continue to play intensely—jockeying for position, competing, and arguing hour after hour about who'll be the boss or who will get access to the coveted toy. If a woman is pushed over the edge by finding out that her husband is having an affair, or if her child is in danger, her anger will blast right through and she will go to the mat. Otherwise, she will avoid anger or confrontation the same way a man will avoid an emotion.

Girls and women may not always feel the initial intense blast of anger directly from the amygdala that men feel. I can remember one time when a colleague did something unfair to me and I came home to tell my husband. He immediately became furious at the person and couldn't understand why I wasn't really mad. Instead of triggering a quick action response in the brain, as it does in males, anger in girls and women moves through the brain's gut feeling, conflict-pain anticipation, and verbal circuits. I had to chew on the incident for a while before responding. Women talk to others first when they are angry at a third person. But scientists speculate that though a woman is slower to act out of anger, once her faster verbal circuits get going, they can cause her to unleash a barrage of angry words that a man can't match. Typical men speak fewer words and have less verbal fluency than women, so they may be handicapped in angry exchanges with women. Men's brain circuits and bodies may readily revert to a physical expression of anger fueled by the frustration of not being able to match women's words.

Often when I see a couple who are not communicating well, the problem is that the man's brain circuits push him frequently and quickly to an angry, aggressive reaction, and the woman feels frightened and shuts down. Ancient wiring is telling her it's dangerous, but she anticipates that if she flees she'll be losing her provider and may have to fend for herself. If a couple remains locked in this Stone Age conflict, there is no chance for resolution. Helping my patients understand that the emotion circuits for anger and safety are different in the male and female brains is often quite helpful.

ANXIETY AND DEPRESSION

Sarah came into the office one day shaking. She and Nick had been fighting over the woman he was flirting with in his office. Sarah was convinced Nick had flirted right in front of her that weekend at a dinner party. Whenever he cut off the discussion and left the room, a videotape seemed to lock Sarah's mind into watching the divorce, the division of assets, and the assignment of child custody; saying goodbye to his family; and leaving town. She was having a hard time focusing; she was on alert for the next fight and was becoming sure that their marriage was collapsing.

It wasn't true. Nick was making a big effort, but the arguing was leaving Sarah's brain in acute neurochemical distress. All her brain circuits were on red alert. Nick seemed unperturbed, playing his regular Wednesday night game of hoops. He didn't seem awkward around her at home, yet she was losing sleep, crying all day, and becoming increasingly hopeless. According to Sarah's reality, the world was coming to an end, but Nick seemed to be showing complete indifference.

Why was Sarah feeling unsafe and afraid while Nick was not? Males and females have different emotional circuitry for safety and fear reinforced by our particular experiences in life. The feeling of safety is built into the brain's wiring, and scans show that girls' and women's brains activate more than men's in anticipation of fear or pain. According to research at Columbia, the brain learns about what is dangerous when its fear pathways are activated and about what is safe when its pleasure-reward circuits fire. Females find it harder than do males to suppress their fear in response to anticipation of danger or pain. This is why Sarah was freaking out at home alone.

Anxiety is a state that occurs when stress or fear triggers the amygdala, causing the brain to rally all its conscious attention to the threat at hand. Anxiety is four times more common in women. A woman's highly responsive stress trigger allows her to become anxious much more quickly than does a man. Although this may not seem like an adaptive trait, it actually allows her brain to focus on the danger at hand and respond quickly to protect her children.

Unfortunately, this intense sensitivity in adult women, as in teenage girls, means that they are nearly twice as likely as men to suffer from depression and anxiety, especially through their reproductive years. This troubling phenomenon exists across cultures, from Europe, North America, and Asia to the Middle East. While psychologists have emphasized cultural and social explanations for this "depression gender gap," more and more neuroscientists are finding that sensitivity to fear, stress, genes, estrogen, progesterone, and innate brain biology play important roles. Many gene variations and brain circuits that are affected by estrogen and serotonin are thought to increase women's risk of depression. The CREB-1 gene, which is different in some women diagnosed with depression, has a little switch that is turned on by estrogen. Scientists speculate that this may be one of several mechanisms by which women's vulnerability to depression turns on at puberty with the surges of progesterone and estrogen. Estrogen's effects may also explain why three times more women than men suffer from the "winter blues," or seasonal affective disorder. Researchers know that estrogen affects the body's circadian rhythm, the sleep and wake cycle stimulated by daylight and darkness, triggering these "wintertime blues" in genetically vulnerable women.

Every year scientists are locating more gene variations related to depression that run in certain families. Another gene, called the serotonin transporter gene—or 5-HTT—also seems to trigger depression in females who inherit a particular version of it. Scientists speculate that this gene variation may contribute to making depression more common in women, because its switch is triggered by threats and severe stress. This may have been the situation in Sarah's case—she came from a family with a history of depression only in the female members. As I know from the many women who come to my clinic, it is often the severe stress caused by the loss of a relationship that pushes genetically vulnerable women over the edge into a clinical depression. Other hormonal events—pregnancy, postpartum depression, premenstrual syndrome, perimenopause—can also disrupt the female brain's emotional balance, and during a rough period a woman may need chemical or hormonal rebalancing.

KNOW THE DIFFERENCE

As both men and women grow into middle and older ages, gain more life experience, and feel more secure, they often become more comfortable expressing a fuller range of emotions, including those—for men especially—they have long suppressed. But there's no getting around the fact that women have different emotional perceptions, realities, responses, and memories than do men, and these differences—based on brain circuitry and function—are at the heart of many interesting misunderstandings. Evan and Jane came to see each other's realities. When she broke down crying out of the blue, he tried to figure out if he was being unresponsive in some way. When she was tired and didn't want to have sex, he fought his instincts and took her at her word. When he became irritable and possessive, she realized she hadn't been sexually attentive enough. And just as they had come to understand each other, it was all about to change. There was still one major shift to come in the female reality.

SEVEN

The Mature Female Brain

SYLVIA WOKE UP one day and decided, this is it. I'm done. I want a divorce. It had become clear to her that her husband, Robert, was unavailable and ungiving. She was tired of listening to his tirades and fed up with his demands. But what really pushed her over the edge was when she found herself in the hospital for a week for an intestinal blockage and he visited her only twice. Both times he came to ask questions about running the house.

At least this is how Sylvia, an attractive woman with brown hair, bright blue eyes, and a spring in her step, explained it to me during a therapy session. Since her early twenties, she felt she had spent most of her time taking care of needy, self-absorbed people. She had fixed their problems, pulling them out of alcoholism or abusive situations, and in return they had sucked her emotionally dry. At age fifty-four, she was still very attractive and felt full of energy. What astounded her more than

179

anything was that she felt as though a haze had lifted recently, and she could see in a way she hadn't been able to before. The tugs she used to feel at her heartstrings to rescue and care for others had all but vanished. She was ready to take some risks and start walking in the direction of her dreams. "What is it about my life that isn't working?" she asked. "I want more out of my life than this!" For years she had cooked and cleaned and raised three children as a stay-at-home mother. Though she had yearned to work, Robert had made it impossible by denying her household help. For twenty-eight years she had chauffeured, nurtured and loved their children, made sure homework was done, dinner was eaten, and the house didn't fall apart. Now, out of nowhere, she found herself asking, Why?

Sylvia's story has become an all too familiar rite of passage: the menopausal woman chucking everything, and everyone, and starting over, especially now that 150,000 American women per month are entering this phase of life. It's a process that seems baffling to the premenopausal woman and has shocked more than a few husbands. A menopausal woman becomes less worried about pleasing others and now wants to please herself. This change has been looked at as a moment of psychological development, but it is also likely triggered by a new biological reality based in the female brain as it makes its last big hormonal change of life.

If we took our MRI scanner into Sylvia's brain, we'd see a landscape quite different from that of a few years before. A constancy in the flow of impulses through her brain circuits has replaced the surges and plunges of estrogen and progesterone caused by the menstrual cycle. Her brain is now a more certain and steady machine. We do not see the hair-trigger circuits in the amygdala that rapidly altered her reality right before her period, sometimes pushing her to see bleakness that wasn't there or to hear an insult that wasn't intended. We would see that the brain circuits between the amygdala (the emotional processor), and the prefrontal cortex (the emotion assessment and judgment area) are fully functional and consistent. They are no longer easily overamped at certain times of the month. The amygdala still lights up more than a man's when Sylvia sees a threatening face or hears about a tragedy, but tears don't flood her so quickly anymore.

Fifty-one and a half years is the average age of menopause, the moment twelve months after a woman's last period; twelve months after the ovaries have stopped producing the hormones that have boosted her communication circuits, emotion circuits, the drive to tend and care, and the urge to avoid conflict at all costs. The circuits are still there, but the fuel for running the highly responsive Maserati engine for tracking the emotions of others has begun to run dry, and this scarcity causes a major shift in how a woman perceives her reality. With her estrogen down, her oxytocin is down, too. She's less interested in the nuances of emotions; she's less concerned about keeping the peace; and she's getting less of a dopamine rush from the things she did before, even talking with her friends. She's not getting the calming oxytocin reward of tending and caring for her little children, so she's less inclined to be as attentive to others' personal needs. This can happen precipitously, and the problem is, Sylvia's family can't see from the outside how her internal rules are being rewritten.

Until menopause, Sylvia's brain, like most women's, has been programmed by the delicate interplay of hormones, physical touch, emotions, and brain circuits to care for, fix, and otherwise help those around her. Societally, she has always been reinforced for pleasing others. The urge to connect, the highly tuned desire and ability to read emotions could sometimes compel her to help even in hopeless cases. She explained to me the times she had chased her friend Marian around town making sure Marian didn't drive when she was out on a bender; Sylvia spent most of her forties trying to please a demanding father, who had become senile after the death of her mother; and she stayed with Robert convinced that if she kept the peace just a little longer, everyone would remain in the family unit and they'd all be okay. Their marriage had never been a strong one. She had always been worried, Sylvia said, when the kids were young, that if Robert and she split something disastrous would happen to the children.

But now that the kids were grown and out of the house, the circuits that had provided the foundation for these impulses were no longer being fueled. Sylvia was changing her mind. She now wanted to help people on a grander scale—outside the family. As one modern role model to middle-age women, Oprah Winfrey, poetically put it after turning fifty,

> I marvel that at this age I still feel myself expanding, reaching out and beyond the boundaries of self to become more enlightened. In my twenties, I thought there was some magical adult age I'd reach (thirty-five, maybe) and my "adultness" would be complete. Funny how that number kept changing over the years, how even at forty, labeled by society as middle-aged, I still felt I wasn't the adult I knew I could be. Now that my life experiences have transcended every dream or expectation I ever imagined, I know for sure that we have to keep transforming ourselves to become who we ought to be.

Once her estrogen level dropped, oxytocin—the connecting and tending hormone—also dropped. Instead of off-the-charts spikes, Sylvia's emotional, tending, and nursing impulses were dialed down to a dull, steady roar. There's a new reality brewing in Sylvia's brain, and it's a take-no-prisoners view.

This has become the twenty-first-century reality of ancient female brain wiring. This changed reality in Sylvia's brain is the foundation of her newly found balance. The brain's circuits don't change all that much in the mature female brain, but the high-test fuel—estrogen—that ignited them and pumped up the neurochemicals and oxytocin in the past has eased off. This biological truth is a powerful stimulus for the road ahead. One of the great mysteries to women at this age—and to the men around them—is how the changes in their hormones affect their thoughts, feelings, and the functioning of their brains.

PERIMENOPAUSE: THE ROCKY BEGINNING

A woman's hormones have been changing for several years before her day of menopause arrives. Starting at about age forty-three, the female brain becomes less sensitive to estrogen, touching off a cascade of symptoms that can vary from month to month and year to year, ranging from hot flashes and joint pain to anxiety and depression. Scientists now believe that menopause is triggered by this change in estrogen sensitivity in the brain itself. Sex drive can change radically, too. The level of estrogen drops, and so does that of testosterone—the rocket fuel for sex drive. The stability of the female brain's reality, in fact, can be an almost daily uncertainty by age forty-seven or forty-eight. The twenty-four months before menopause, while the ovaries make erratic amounts of estrogen before stopping production of the hormone altogether, can be a rocky ride for some women.

That's how Sylvia felt at age forty-seven, when she called my clinic to make an appointment—the first time in her life she had seen a psychiatrist. It was the year before her last child left for college, and she had constant mood symptoms—including irritability, with emotional outbursts and a lack of joy or hope—that had started to distress her. "Perimenopause is like adolescence—without the fun," she said one day. It's true: your brain is at the mercy of changing hormones, as it was in puberty, with all the nerve-jangling psychological stress responsivity, worries about appearance, and over-the-top emotional responses. Sylvia would be fine one minute, but just the wrong comment from Robert could send her slamming doors throughout the house and taking refuge in the garage for an hour-long sob fest. She couldn't take it anymore and wanted me to prescribe something to treat her symptoms. The other issues with Robert would have to wait. So I gave her estrogen and Zoloft. In two weeks she was amazed at how much better she felt. Her brain needed the neurochemical support.

For a lucky 15 percent of women, the perimenopause—the two to nine years before menopause—is a breeze, but for about 30 percent it can cause major discomfort, and 50 to 60 percent of women experience some perimenopausal symptoms at least some of the time. Unfortunately, there's no way to know how you'll react until you get there.

There are some clear signs, however, when you've crossed the threshold. Your first hot flash, for one thing, is a signal that your brain's starting to experience estrogen withdrawal. Your hypothalamus, in response to decreased estrogen, has changed its heat-regulating cells, making you feel suddenly, blazingly hot even at normal temperatures. Another sign of perimenopause is the shortening of your menstrual cycle by a day or two, even before you've experienced your first hot flash. The brain's response to glucose changes dramatically, too, giving you energy surges and drops and cravings for sweets and carbs. This estrogen withdrawal affects the pituitary, curtailing the menstrual cycle and making the timing of ovulation and fertility unreliable. So be careful—many women end up with a surprise "change-of-life" baby thanks to the breakdown in the predictability of their ovulation.

I started the Women's Mood and Hormone Clinic long before I was in perimenopause or menopause, so all I personally had experienced was moderately bad PMS and postpartum hypothyroidism. But when I was in my mid-forties, I began to have extremely bad PMS, with high irritability and big mood drops. At first I thought it was the stress of my job and having primary responsibility for my son. No doubt those realities played into my perimenopausal syndrome, but I resisted taking hormones for several years, thinking, Oh, this isn't the same thing I see in my patients every day. Boy was I wrong. By the age of forty-seven, I was in full-blown perimenopause. I couldn't sleep well, woke up hot, and often had to change my nightgown. In the morning I felt like hell: tired, irritable, and ready to cry over anything. Two weeks after starting estrogen and Zoloft, I miraculously felt like my old self again.

Since estrogen also affects the brain's levels of serotonin, dopamine, norepinephrine, and acetylcholine—neurotransmitters that control mood and memory—it's no surprise that big changes in estrogen level can influence a wide variety of brain functions. This is where drugs such as Zoloft and other SSRIs can help, because they prop up these neurotransmitters in the brain. Studies show that perimenopausal women complain of more symptoms of all kinds—from depressed mood and sleeping problems to memory lapses and irritability—to their doctors than do women who have already passed through menopause. Interest, or the lack of it, in sex can be an issue as well. Along with the estrogen drop, testosterone—the fuel for love—can plummet at this time too.

THE WOMAN'S LAST GYNO-CRISIS

Marilyn and her husband, Steve, came to see me when Steve was at his wit's end from being rejected sexually by her. "She won't let me touch her anymore," he said. Marilyn told me, "I used to like sex a lot and would like to have that feeling again, but every time he touches me, or gets that look in his eye, it's . . . it's . . . just irritating. It's not that I don't love him. I do." Husbands can feel dumbfounded—a man's hormones haven't abruptly changed—even though they will decrease, and gradually he, too, will have fewer sexual urges. But his brain will never go through the precipitous hormone decline a woman's brain has had to endure.

It was a good thing they came in, since this was a biological problem that was quickly becoming a marital problem. Many women do experience a drop in libido, but I suspected Marilyn's perimenopausal situation was a little more extreme than most. I measured her testosterone and found that it was barely present. Could this be the cause of her rejection of Steve? She decided to find out by trying testosterone, so I prescribed the patch, and she slapped it on that very day.

Although sexual response varies a great deal during these years of erratic hormones, 50 percent of women age forty-two to fifty-two lose their interest in sex, are harder to arouse, and find their orgasms are much less frequent and intense. By the age of menopause, women have also lost up to 60 percent of the testosterone they had at age twenty. But this can be replaced by many forms of testosterone supplements—such as patches, pills, and gels—that are now available.

When I greeted Marilyn and Steve in the waiting room two weeks later, Steve gave me two thumbs up. Marilyn said that within a week she began to feel less annoyed at his sexual advances, and during the second week, she even felt like initiating sex herself but didn't. Her brain circuits for sexual desire had been reignited by a little hormonal rocket fuel. Use it or lose it goes for everything, memory and sex included. The brain below the waist will shrivel up if it isn't used.

Not all perimenopausal or postmenopausal women lose their testosterone or their sexual interest. In fact, "postmenopausal zest" is a phrase coined by the anthropologist Margaret Mead. It is a time when we no longer have to be concerned with birth control, PMS, painful cramps, or other monthly gynecological inconveniences. It is a stage of life that is free from many en-cumbrances and full of wonderful possibilities. We are still young enough to live life to its fullest and enjoy all the good things nature has provided us. Many women experience a renewed zest for life, even rejuvenated sexual desire, and look for exhilarating adventures or new beginnings. It is like starting life all over with a better set of rules. For those who don't have the zest, the testosterone patch may ignite it.

By the time Sylvia decided to see me again about divorcing Robert—after he hadn't come to visit her much in the hospital—she had passed through the last throes of peri-menopause and stopped taking the estrogen and Zoloft. It was then that Sylvia explained to me it felt as if a haze had lifted in her brain once her menstrual cycle had stopped. She had always suffered terribly from PMS, and now that that was over, it was as if her vision had become clearer—about what she wanted to do with her life, and what she didn't want to do any longer. She told Robert that although she still respected him, she had grown tired of demands that she continue to tend his needs on his timetable and keep up their large home. The monthly priming of her brain circuits by surging estrogen and oxytocin—to assure that she would tend to the needs of others—was gone. Of course she still had that blistering love for her children, but she didn't have their physical presence and their oxytocin-stimulating hugs or her estrogen pulses to trigger her caretaking circuits and behaviors anymore. Of course she could still perform these duties, but she no longer felt driven to. She turned to Robert and said, "You're a grown-up and I'm finished raising the kids. Now it's my turn to have a life."

When her kids came home during a break from college, Sylvia reported that she really enjoyed seeing them and catching up on their lives but was annoyed that they still expected her to pick up after them, cook their meals, and do their laundry. Her kids even teased her about how she would throw their laundry in the washer and dryer but wouldn't match their socks anymore. She had laughed, too, but for the first time in her life, she said, she blasted back a reply: "Do your own damn laundry, it's about time you grew up!"

The mommy brain was beginning to unplug. When a woman has launched all her children, her ancient mommy wiring comes loose and she is allowed to pull a few of the connections to the child-tracking device out of her brain. When the umbilical cord is cut as the children leave home, the mommy brain circuits are finally free to be applied to new ambitions, new thoughts, new ideas. Many women, however, may feel desperately sad and disoriented when their children first leave home. These circuits, which evolved for millions of years in our foremothers, fueled by estrogen and reinforced by oxytocin and dopamine, are now set free.

This time of life for some women is not so rancorous as it was for Sylvia. My patient Lynn had a deep and loving marriage to Don for over thirty years by the time their two kids were on their own at college. Lynn and Don started to travel to the places they had always wanted to go. They felt a sense of satisfaction at having raised two wonderful and accomplished children. Lynn had enjoyed being a mother but found that after a few months of heart tugs when they went off to college, she enjoyed not having to deal with the morning routine of getting the kids out the door. She was a successful—and well-liked—administrator at the university. Don was an engineer in private industry. The more time they spent alone, the more their relationship blossomed. They brought years of mutual love and trust to help them through this life transition and set up the new rules for the road ahead.

Sylvia's midlife transition was not nearly so peaceful. By our next session, she had decided to go back to grad school and begin working in a mental health clinic two times a week. Her kids were a little unsettled by her new interests. The youngest was moving on and getting settled into life at college. She didn't need her mother as much as she used to, but still, she was surprised and a little hurt when she talked to Sylvia on the phone and all her mother wanted to tell her was about her own new projects and plans for going back to school herself. Sylvia told me that she almost found it shocking that she was no longer anxiously asking her daughter questions about her life. She was amazed at her slightly detached response.

What's happening in her brain? It isn't just that the estrogen is gone—the physical sensations of tending and touching the children are also gone. Those sensations, along with estrogen, help to reinforce the tending circuits and turn up oxytocin in the brain. This process begins for most mothers during their children's teen years, when they resist being hugged, kissed, or touched. So by the time they leave the nest, mothers have grown accustomed to less up close and personal physical tending. An experiment on mothering behavior in rats found that physical contact is required to maintain the female brain circuits for active maternal behavior. Scientists numbed the chest, abdomen, and nipple area on the rats. The mothers could see, smell, and hear their pups but could not feel them squirming around. The result: mothering and bonding behaviors were severely impaired. The mothers didn't fetch, lick, and nurse their pups the way a normal rat mom would. Even though their brain circuits were organized and primed hormonally for mothering and caretaking behaviors, without the feedback of touch sensation, the mother rats' brain connections for nurturing behavior did not develop, and many of the pups died as a result.

Human mothers also use this physical feedback to activate and maintain nurturing and caretaking brain circuits. The normal contact of living in the same house provides enough sensation to maintain a woman's tending and caretaking behaviors toward her kids—even grown-up kids. Once the kids leave the house, however, that's another story. If a mother is menopausal at the same time, the hormones that built, primed, and maintained those brain circuits are also gone.

This change doesn't mean that the tending brain circuits are gone forever. Four out of five women over fifty say having a job where they help others is important to them. Though the initial impulse for many menopausal women seems to be doing something for themselves for once, the renewal that follows often draws them back toward helping others. The caregiving circuits can easily be renewed. If an over-fifty-year-old woman becomes the mother of a new baby, the daily physical contact will cause those circuits to reemerge in her brain—as one of my female colleagues could tell you after she adopted a Chinese baby girl when she was fifty-five years old. So once the circuits are there, they can be reignited. It isn't over until it's over as far as the maternal female brain is concerned.

For Sylvia, though, this was a golden time. In her reality, she was free at last to follow her own pied piper. She had taken on her own projects. Through her new courses, she had become convinced that behavioral problems in teens have their roots in early education, and she became passionate about improving how parents and teachers treat preschoolers. As part of getting her master's in social work, she became involved in training preschool teachers in the local school system. She told me that she had also returned to services at the church where she grew up and was building a studio in her garage so that she could go back to painting—an activity she had given up when she married Robert. At one of our sessions, she was almost in tears over how happy her new life was making her. She felt she was making a difference in the world. This was in direct contrast to the increasingly heated arguments that began the minute Robert walked in the door every night.

WHO ARE YOU AND WHAT HAVE YOU DONE WITH MY WIFE?

Soon Sylvia and Robert came to see me together for another couple's session. Unresolved issues for both of them had finally come to a head. Robert couldn't believe what he was hearing. For instance, "Make your own damn dinner or go out by yourself. For the last time, I'm not hungry. I'm happy painting right now and I don't feel like stopping." He said she had snapped at him at a party two nights before when she offered a suggestion about investing in a group of stocks and he told her to stay out of the discussion because she didn't know what she was talking about. He was the one who read Barron's, after all. "Yeah, you keep reading it, and you keep losing money. Have you seen my portfolio lately? I've made three times the amount you've made, so stop belittling me," she'd replied. Everything he said seemed to annoy her. She announced she was moving out.

When Sylvia was younger, she would do everything she could to avoid fights with her husband, even if she was really mad. Remember the tape that gets rolling during the teen years, when estrogen dials up the emotions and communication circuits—the one that makes a woman panic about any conflict as a threat to a relationship? That tape doesn't stop rolling until a woman either consciously overrides it, or the supply of hormones that fuels it is cut off, or both. A time like now. All her life Sylvia had prided herself on being coy, accommodating, and willing to let her husband win—especially when he came home exhausted and on edge from the office. Her empathy for him was real. She kept the peace, as her Stone Age brain was compelling her to do, to keep the family together. Having a husband is good. We're better protected this way. These were the messages keeping her from engaging in conflict. If Robert forgot their anniversary, she would bite her tongue. If he was verbally abusive after a long day at work, she stared straight into the stew she was stirring and didn't respond.

But as Sylvia hit menopause, the filters came off, her irritability increased, and her anger wasn't headed for that extra "stomach" anymore, to be chewed over before it came out. Her ratio of testosterone to estrogen was shifting, and her anger pathways were becoming more like a man's. The calming effects of progesterone and oxytocin weren't there to cool off the anger either. The couple had never learned to process and resolve their disagreements. Now Sylvia confronted Robert with regularity, venting decades of pent-up rage.

At their next session, it became clear that it was not all Robert's fault. He was going through his own, more modest, life changes. But Sylvia still wanted to move out. Neither of them was yet aware of the changing reality in her brain, which was rewriting the rules not just for arguing but for every interaction of their relationship. Studies show that women who are unhappy with their marriages report more negative moods and illnesses during the menopause years. So when the hormonal haze lifts and the children leave home, women often find themselves more unhappy than they could allow themselves to realize before. Often all the unhappiness gets blamed on the husband. Obviously, Sylvia had her legitimate complaints about Robert. But the root cause of her unhappiness was still unclear.

The next week she reported that her daughter had said, "Mom, you're acting weird, and Dad is getting scared. He says you're just not the woman he's been married to for nearly thirty years, and he's afraid you'll do something crazy—like take all the money and run away." Sylvia wasn't crazy, and she wasn't going to abscond with their savings, but it was true, she wasn't the same woman. She told me that her husband once screamed at her, "What have you done with my wife?" A huge number of her brain circuits had been abruptly shut down, and just as abruptly, Sylvia had changed the rules of their relationship. As often happens in these situations, nobody told Robert.

It is commonly believed that men leave their aging, chubby, postmenopausal wives for fertile, younger, thin women. This couldn't be further from the truth. Statistics show that more than 65 percent of divorces after the age of fifty are initiated by women. My suspicion is that much of this female-initiated divorce is rooted in the drastically altered reality of postmenopausal women. (But as I have seen in my practice, it could also be because they are tired of putting up with difficult or cheating husbands and have just been waiting for the day when the children leave home.) What had been important to women—connection, approval, children, and making sure the family stayed together—is no longer the first thing on their minds. And the changing chemistry of women's brains is responsible for the shifting reality of their lives.

During any time when hormones are shifting and hijacking your reality, it's important to examine impulses and make sure they're real, as opposed to hormone-induced. Just as the drops in estrogen and progesterone before a period can make you believe you're fat, ugly, and worthless, the absence of reproductive hormones can make you believe your husband is the cause of all your misery. Maybe he is. And maybe he isn't. As Sylvia learned through our discussion, if you understand some of the biological reasons for your changing feelings and reality, you might just learn to talk about it with him—and he might just change. It's a long process of education, one that best begins before the "change" takes place.

WHO'S COOKING DINNER?

During our session after my August vacation, Sylvia told me that she had decided she wanted a divorce after all. As a matter of fact, she had moved out the month I was away. Her friends had even started setting her up with new men. It didn't take long before she was as annoyed with them as she was with Robert. Sylvia quickly discovered that older men were looking for a "nurse with a purse"—someone who had her own money and would take care of them for the rest of their lives. This was a bit shocking to her. It was just what she had been looking for in a man when she was young. Back then she wanted someone who would take care of her and bring the money with him. Back then she was willing to take care of him along with the children. Now it was the last thing on her mind.

Sylvia still felt hopeful that she would find the "perfect man" to grow old with, an equal partner, a soul mate, someone she could talk to and share life's joys with, but not do the physical caretaking, shopping, cooking, laundry, and cleaning that many of the men she was dating had come to expect of their ex-wives. As she put it, she had no intention of being a nurse, and she didn't want someone to steal her purse. "Otherwise," she said, "I'd rather have no one right now." After all, she had lots of dear friends who made her happy. She was looking forward to a much less psychologically stressful existence than what she had been experiencing lately in arguing with Robert.

This decreased urge to tend and nurture after menopause may not come as a relief to all women. Research has yet to examine the effects of low oxytocin, which ensues after estrogen declines, but it may lead to some real behavioral changes. Most women, however, are only vaguely aware of it—if at all. My patient Marcia, age sixty-one, admitted to me, for example, that she was feeling much less concerned about the problems and needs of her family, friends, and children, and less inclined to look after them. Nobody had complained to her about this decreased caretaking, though her husband wondered why he'd been fixing his own dinner a lot. Mainly, it was just something Marcia noticed in herself. She didn't really mind her newfound emotional independence—she was spending more time on solo pleasures, such as the genealogical research she loved to do. She had not had a menstrual period in over four years. But her vaginal dryness, night sweats, and interrupted sleep had led her to start being treated with estrogen pills. Three months after she began estrogen therapy, however, Marcia's nurturing instincts had returned. She hadn't recognized how drastically they had changed over the past four years until they came flooding back. She told me she was shocked that one little pill could make her feel more like her old self—a self she only vaguely realized she had lost. Estrogen therapy may have stimulated her brain to produce higher levels of oxytocin again, triggering familiar, affiliative patterns of behavior, to her husband's relief.

THE LAST TIME a woman had a nonfluctuating stress responsivity because of steady, low hormones was in the juvenile pause, or during the months of pregnancy, when the pulsing cells of the hypothalamus are shut down and the stress response is kept low. After ten years of no hormones, one of my post-menopausal patients reported to me that, although her sex drive was suffering, she and her husband had stopped fighting when they went on trips. It used to be that travel really stressed her out, but suddenly she was loving every minute of waking up early to catch a plane and going to unknown places. She even liked packing, and as the stress faded away, their travel fights had disappeared as well.

As for Sylvia, soon after she moved out, she noticed that her mood swings and irritability stopped. She told me that her work with preschool teachers and parents had allowed her to become the person she always knew she ought to be. She began to look forward to the nights she spent alone, watching old movies, taking long bubble baths, and working late in her new studio. If her kids called, she was always eager to talk to them, but she found that she would not become as engaged in helping to solve their problems, getting upset, or giving them endless advice. At first she thought the reason her moodiness and irritability had decreased was that she had gotten the biggest problem out of her life: her bad marriage. But she had also noticed that her hot flashes had almost disappeared and she was sleeping well again.

When she came to see me six months after leaving Robert, I gently queried whether it was only that her husband was out of the house, or whether it might also be that she had now settled into a new hormonal state, in which her mood was steadier. Sylvia also mentioned that she was less irritable, and during this session she even complained about being lonely and having nobody with whom to discuss the events in her children's lives and her own. I suggested that she might be missing Robert's company and that if they started spending time together but negotiated a new set of rules, she might notice that their relationship was more balanced.

Just Getting Started

At menopause, the female brain is nowhere near ready to retire. As a matter of fact, many women's lives are just hitting their peak. This can be an exciting intellectual time now that the burden of rearing children has decreased and the preoccupation of the mommy brain is lessened. The contribution of work to a woman's personality, identity, and fulfillment once again becomes as important as it may have been before the mommy brain took over. When Sylvia found out that she was accepted into a master's program in social work, it was one of the happiest days of her life. She hadn't had such a feeling of accomplishment since she graduated from college, got married, or had her children.

As a matter of fact, work and accomplishment can be critical to a woman's sense of well-being during this life transition. Studies have shown that women with high career momentum at this stage of life viewed their work as more central to their identities than did women who were just maintaining or decreasing their career momentum. Also, women with high career momentum scored better on measures of self-acceptance, independence, and effective functioning in their fifties and sixties, and rated their physical health higher than did other women. There's a lot of life left after menopause, and embracing work—whatever that may be—passionately clearly allows a woman to feel regenerated and fulfilled.

Leave Me Alone Already

Edith made an appointment with me as her husband, a psychiatrist, was winding down his practice in order to retire.

Although they had a good relationship most of the time, all she could envision was that he would be constantly invading her space, demanding she serve him twenty-four hours a day. Her distress over the idea had given her insomnia. And she turned out to be right. The minute he was home, he started asking, "Where's lunch? Did you buy my salami? Who moved my tool-box? Aren't you going to clean up the dishes? They've been sitting in the sink for an hour." When she hadn't gone shopping because she was busy, he said, "Busy with what?" She had been helping her mother's oldest friend with things around the house. She had been taking care of her grandchildren on Tuesdays. She had a regular bridge game and lunch dates, and she attended a book club. She was busy working at the things that mattered to her. She liked her freedom. Her husband was dumbfounded that she showed little interest in him and had so much of her own life to live.

This change in behavior is actually the most common one I see in women sixty-five and older. Like Edith, they come into my office depressed, anxious, and unable to sleep. I soon find out that their husbands have retired over the past year. They feel conflicted, angry, and pulled away from their own work and activities. They don't want to live this way for the rest of their lives. This fear of losing freedom can happen even if the marriage relationship is basically good. Somehow many women feel that they cannot renegotiate the unwritten marriage con-tract. "Of course you can," I tell them. "Your life depends on it."

Weeks later, after Edith and her husband had been on a month's vacation, she returned to see me. A pleased grin on her face, she said, "Mission accomplished! He has agreed to keep out of my hair." They had renegotiated the rules for the next phase of their life.

Hormones in the Female Brain After Menopause

Hormones in the brain are part of what makes us women. They are the fuels that activate our sex-specific brain circuits, resulting in female-typical behavior and skills. What happens to our female brains at menopause, when we lose this hormonal fuel? The brain cells, circuits, and neurochemicals that have relied on estrogen soon shrivel. In Canada the researcher Barbara Sherwin found that women who had estrogen replacement therapy right after the removal of their ovaries retained the memory function they'd had before, but women who had no estrogen replacement right after their ovaries were removed had declining verbal memory unless they were soon given estrogen. The therapy restored their memory to nearly premenopausal levels—but only if they began it immediately or soon after the operation. There's a brief window, it seems, when estrogen provides maximum protective benefits for the brain.

Estrogen may have a protective effect on many aspects of brain functioning, even on the mitochondria—energy centers of cells—especially those in blood vessels in the brain. Researchers at the University of California, Irvine, found that estrogen treatment increased the efficiency of these mitochondria, perhaps explaining why premenopausal women have lower rates of stroke than men their age. Estrogen can help the brain's blood flow stay strong for years into older age. At Yale University, for instance, researchers treated postmenopausal women with estrogen or a placebo for twenty-one days, then scanned their brains while they performed memory tasks. The women on estrogen had brain patterns characteristic of younger subjects, while those without estrogen had brain patterns typical of much older women. And yet another study, of brain volume in postmenopausal women, suggested that estrogen protects specific parts of the brain. In women who took estrogen, there was less shrinkage in the brain areas for decision making, judgment, concentration, verbal processing, listening skills, and emotional processing.

The protective effect estrogen appears to have on female brain function is one reason scientists are carefully reconsidering the results of the Women's Health Initiative in 2002, a study that found that women who started taking estrogen after a thirteen-year gap post-menopause didn't get its protective effects on the brain. Scientists have now shown that a gap of more than five or six years after menopause without estrogen means that the opportunity to reap estrogen's preventative effects on the heart, brain, and blood vessels is likely gone. Early treatment with estrogen may be especially important to protect brain function as well.

Many women have felt confused and betrayed by the fact that they were told one thing a few years ago by their doctors about hormone replacement therapy (HRT), now called HT, but now hear the opposite based on the results of the WHI study. I myself—as both a doctor and a postmenopausal woman—have been caught in that bind. How and when to start HT and when and if to stop remain burning questions for patients and doctors alike. Until new studies clarify this issue, however, each patient must find her own way—using diet, hormones, activities, exercise, appropriate treatment, and regular input from informed doctors who are specialists in hormone therapy. I now have a complete discussion with each of my menopausal patients about her family genetics, lifestyle, symptoms, health issues, and the risks and benefits of HT for her.

Despite the storms and hormonal adjustments of menopause,

most women stay remarkably vigorous, smart, and capable as they age, even without the assistance of estrogen. Not all women need or want hormone therapy. It's usually not until decades after menopause that the natural process of aging starts affecting the functioning of the female brain. Men's and women's brains age differently, with men losing more of the cortex sooner than women.

While every woman's body and brain react differently in the years after menopause, for many this is a time of increasing freedom and control over our lives. Impulses are less likely to confuse or agitate us. Our survival may no longer depend on a steady paycheck, and there's less value in pretending about how we feel and more in presenting and living our passionate, real selves. Helping others and being engaged in solving serious problems in the world can energize us. This is also a time when grandmothering can bring new, often uncomplicated joy. Maybe life does save some of the best for last. My sixty-year-old patient Denise, for example, had always been an independent woman focused on her marketing career, even while she was raising her two children. When her daughter gave birth for the first time, Denise was unprepared, she told me, for the waves of love she felt for her grandchild. "I was completely swept off my feet," she said, "which I never, ever expected. I've got a million things going on in my life, but for some reason I can't get enough of this baby. And my daughter's letting me into her life in a way that she never has before. She needs me now, and I want to be there for her."

The special, supportive role that grandmothers play may be one of the reasons that evolution engineered women to live for decades after they can no longer bear children. Grandmothers, according to the University of Utah anthropologist Kristen

Hawkes, may actually be one of the keys to growth and survival in ancient human populations. Hawkes argues that in the Stone Age, the extra food-gathering efforts of able-bodied post-menopausal women increased the survival rate of young grandchildren. Grandmothers' provisioning and help also enabled younger women to produce more children at shorter intervals, increasing the population's fertility and reproductive success. Even though the life span in hunter-gathering societies is typically less than forty, about a third of all adult women survive past that age, and many go on to live productively into their sixties and seventies. Among the Hadza hunter-gatherer population in Tanzania, Hawkes found, for example, that hard-working grandmothers in their sixties spent more time foraging than did younger mothers, providing food for their grand-children and increasing their chances of survival. Researchers have found similar positive effects of grandmothers among Hungarian gypsies and populations in India and Africa. In rural Gambia, in fact, anthropologists found that the presence of a grandmother improves a child's prospects for survival much more than the presence of a father. In other words, women at menopause, the world over, have the option to embrace the life-sustaining role of grandmother, too.

NOW WHAT DO I DO?

A century ago, menopause was relatively rare. Even in the late nineteenth and early twentieth centuries, the average age of death for women in the United States was forty-nine—two years before the typical woman ends her menstrual cycle. Women in the United States can now expect to live many decades after their periods stop. Science, however, hasn't fully caught up with this change in demographics. Our knowledge about menopause is relatively new and incomplete, though it's advancing rapidly as large populations of women are moving through this once rare transition. Forty-five million American women are now between ages forty and sixty.

Planning for the many years after menopause is historically a new option for women. Being able to visualize exciting projects of their own choosing can be one of the most delightful parts of women's lives in the new century. They may have attained personal and economic power by this time. They may have a broad knowledge base, and for the first time in their lives they may have more exciting options than they ever imagined possible. One scientist friend of mine, Cynthia Kenyon, an expert in aging, believes that in the future women will likely live to be more than 120, a lot of years to imagine.

For Sylvia, imagining her postmenopausal years meant rediscovering Robert. When she came to see me again two years after she and Robert broke up, she told me that after she got back to the girl she once was, felt the joy of rediscovering who she really was, and had dated enough disillusioning older men, she realized she missed Robert. He was the only one she could talk to about certain things—including their wonderful children. One day he invited her to dinner and she decided to accept. They met at a romantic restaurant, talked calmly about what had gone wrong, and ended up apologizing for the unhappiness they had caused each other. They also had new experiences to share—her job, her painting, his new interest in antiques, and even their funny adventures in dating. Over time they rediscovered their friendship and respect for each other and realized that they had already found their soul mates. They just needed to rewrite the contract.

The Mature Female Brain

* * *

THE MATURE FEMALE brain is still relatively unknown territory, but it's a wide open place for women to discover, create, contribute, and lead in positive ways for future generations. And maybe even have the most fun years of their lives. The postmenopausal years can be a time for both men and women to redefine their relationships and roles, and take on new challenges and adventures independently and together.

I know for myself that having raised my son, discovered passion in my work, and finally having found my soul mate makes me feel very grateful for my life. The struggles along the way have certainly been painful as well, but they have also been my greatest teachers. The reason I wrote this book was to share my knowledge about the inner workings of the female brain with other women who are traveling their own similar paths, trying to be true to themselves and understand how their innate biology affects their reality. I know it would have helped me to know more about what my brain was doing during many of the craziest times in my life. At each step of the way we can better understand our world if we can have a vision of what our brains are doing. Learning how to harness the female brainpower we have will help us each become the woman we ought to be. As a postmenopausal woman, I find myself excited and more determined than ever to try to make a difference in the lives of the girls and women I touch. Of course, I still can't see around the corner for myself—but the many decades ahead seem full of hope, passion, and outward momentum. I hope this map will help guide you through the incredible journey of the female brain.

The Future of the Female Brain

I F I HAD to impart one lesson to women that I learned through writing this book, it would be that understanding our innate biology empowers us to better plan our future. Now that so many women have gained control over their fertility and achieved economic independence, we can create a blueprint for the road ahead. That means making revolutionary changes in society and our personal choices of partners, careers, and the timing of our children.

Since women are now taking their twenties to get educated and establish their careers, more career women are pushing the boundaries of their biological clocks and having children in their mid- to late thirties—even early forties. A large percentage of my residents in their mid-thirties haven't yet even found the men with whom they want to start families because they have been so busy building their careers. This doesn't mean that women have made bad choices. It means that the phases of women's lives have drastically expanded. In early modern Europe, women became fertile at age sixteen or seventeen and had all their children by the time they were in their late twenties. Now, by the time the mommy brain takes over, women are fully entrenched in careers, and that means an inevitable tug-of-war because of overloaded brain circuits. Then women find themselves facing the ups and downs of perimenopause and menopause with toddlers and preschoolers running around the house. At the same time, they are managing busy careers. If a woman hasn't come to see me in her mid-thirties to talk about the challenges of her fertility and career, then she will come in her mid-forties, saying that she just doesn't have time for perimenopause. She can't afford to lose her memory and focus from moods that make her miserable because her hormones are out of sync.

What does all this mean in terms of women's innate brain biology? It doesn't mean that women should get off the path of motherhood combined with career; it just means they may benefit from getting a glimpse of all the balls they will need to juggle starting in their teen years. Obviously, there is no way that any of us can see around the corners in our lives and anticipate every type of support we will need. Understanding what is happening in our brains at each phase, however, is an important first step in controlling our destiny. Our modern challenge is to help society better support our natural female abilities and needs.

My intentions for this book were to help women through the various shifts in their lives: shifts so big they actually create changes in a woman's perception of reality, her values, and what she pays attention to. If we can understand how our lives are shaped by our brain chemistry, then maybe we can better see the road ahead. It's important to visualize and plan for what's coming. I hope this book has made a contribution to the mapping of the female reality.

There are those who wish there were no differences between men and women. In the 1970s at the University of California, Berkeley, the buzzword among young women was "mandatory unisex," which meant that it was politically incorrect even to mention sex difference. There are still those who believe that for women to become equal, unisex must be the norm. The biological reality, however, is that there is no unisex brain. The fear of discrimination based on difference runs deep, and for many years assumptions about sex differences went scientifically unexamined for fear that women wouldn't be able to claim equality with men. But pretending that women and men are the same, while doing a disservice to both men and women, ultimately hurts women. Perpetuating the myth of the male norm means ignoring women's real, biological differences in severity, susceptibility, and treatment of disease. It also ignores the different ways that they process thoughts and therefore perceive what is important.

Assuming the male norm also means undervaluing the powerful, sex-specific strengths and talents of the female brain. Until now, women have had to do most of the cultural and linguistic accommodating in the work world. We have been fighting to adapt to a man's world—after all, women's brains are wired to be good at changing. I hope this book has been a guide—for us, our husbands, fathers, sons, male colleagues, and friends—to the minds and biobehavior of women. Perhaps this information will help men begin adapting to our world.

Almost every woman I have seen in my office, when asked what would be her top three wishes if her fairy godmother could wave her magic wand and grant them, says, "Joy in my life, a

fulfilling relationship, and less stress with more personal time."
Our modern life—the double shift of career and primary
responsibility for the household and family—has made these
goals particularly difficult to achieve. We are stressed out by this
arrangement, and our leading cause of depression and anxiety is
stress. One of the great mysteries of our lives is why we as
women are so devoted to this current social contract, which
often operates against the natural wiring of our female brains
and biological reality.

During the 1990s and the early part of this millennium, a new
set of scientific facts and ideas about the female brain has been
unfolding. These biological truths have become a powerful
stimulus for the reconsideration of a woman's social contract. In
writing this book I have struggled with two voices in my head—
one is the scientific truth, the other is political correctness. I
have chosen to emphasize scientific truth over political correctness even though scientific truths may not always be welcome.

I have met thousands of women during the years my clinic has
been running. They have talked to me about the most intimate
details of events in their childhoods, teen years, career decisions,
choice of a mate, sex, motherhood, and menopause. While female
brain wiring has not changed much in a million years, the
modern challenges of the different phases of women's lives are
remarkably different from those of our foremothers.

Even though there are now proven scientific differences
between men's and women's brains, this, in many ways, is the
Periclean golden age for women. The age of Aristotle, Socrates,
and Plato was the first time in Western history that men gained
enough resources to have the leisure for intellectual and
scientific pursuits. The twenty-first century is the first time in
history that women are in a similar position. We have not only

the critical, unprecedented control over our fertility but independent economic means in a networked economy. Scientific advances in women's fertility have given us enormous options. We can now choose when, if, and how to bear children over many more years of our lives. We are no longer economically dependent on men, and technology has provided the flexibility to toggle between professional and domestic duties at the same time and in the same place. These options give women the gift of using their female brains to create a new paradigm for the way they manage their professional, reproductive, and personal lives.

We are living in the midst of a revolution in consciousness about women's biological reality that will transform human society. I cannot predict the exact nature of the change, but I suspect it will be a shift from simplistic to deep thinking about the changes we need to make on a grand scale. If external reality is the sum total of the way people conceive it, then our external reality will change only when the dominant view of it changes. The scientific facts behind how the female brain functions, perceives reality, responds to emotions, reads emotions in others, and nurtures and cares for others are women's reality. Their needs for functioning at their full potential and using the innate talents of the female brain are becoming clear scientifically. Women have a biological imperative for insisting that a new social contract take them and their needs into account. Our future, and our children's future, depends on it.

The Female Brain
and Hormone Therapy

IN 2002 THE Women's Health Initiative (WHI) and Women's Health Initiative Memory Studies (WHIMS) found that women who took a specific type of hormone therapy for six years, starting at age sixty-four or older, had a small increase in the risk of breast cancer, stroke, and dementia. Ever since, hormone therapy (HT) for women has been downright confusing. Doctors have been massively backpedaling on what they had told their women patients about hormone therapy. And both the doctors and the women caught in the middle have felt betrayed.

The big question remains: whether or not to take hormones during or after menopause. Women want to know, Will the benefits outweigh the risks for me personally? Since the average woman in the WHI study was sixty-four years old and hadn't been on any hormones for thirteen years after menopause, do the study results pertain to, say, a fifty-one-year-old woman now going through menopause and feeling miserable? Or a sixty-something woman who has been on and off hormone therapy? Women ask, Will my brain be able to adjust to no estrogen? Will my brain cells be unprotected if I don't take hormone therapy?

Since the WHI study was not designed to answer questions about hormone therapy and protection of the female brain, we must turn to other studies that have looked directly at the effects of estrogen on the brain.

Estrogen's effect on brain cells and function has been extensively studied in female laboratory rodents and primates. These studies have clearly shown that estrogen promotes brain cell survival, growth, and regeneration. Other studies in women suggest many benefits of estrogen on the growth of neurons and maintenance of brain function as we age. These studies scanned the brains of postmenopausal women, some who took HT and others who did not. The following areas were spared the usual age-related shrinkage in women taking HT: the prefrontal cortex (an area for decision making and judgment), the parietal cortex (an area for verbal processing and listening skills), and the temporal lobe (an area for some emotional processing). Given these positive studies, many scientists now believe that HT should be thought of as a protector against age-related brain decline, although this belief conflicts with the findings of WHI and WHIMS.

It's important to note that there has been no long-term study of the brain effects of estrogen therapy in women who start taking hormones right at menopause, around age fifty-one. The Kronos Early Estrogen Prevention Study, begun in 2005, was designed by Fred Naftolin and colleagues at Yale to research the effects of giving HT to women ages forty-two to fifty-eight, right at the perimenopause and menopause. Its results are due sometime after 2010. Until then, what information other than WHI and WHIMS can we rely on to make our decisions?

On the positive side, the Baltimore Longitudinal Study of Aging—the longest-running scientific study of human aging in the United States, begun in 1958—found numerous brain benefits from HT. Women on hormone therapy, the study shows, have greater relative blood flow in the hippocampus and other brain areas related to verbal memory. They also perform better on verbal and visual memory tests than women who had never been treated with HT. Hormone therapy—with and without progesterone—also helps protect the structural integrity of brain tissue, preventing the usual shrinkage seen with age.

Certain brain regions age faster or more slowly in males and females, just as they develop at different rates early in life. We know that men's brains shrink faster with age than women's brains. This is especially true in regions such as the hippocampus; the prefrontal white matter, which speeds decision making; and the fusiform gyrus, an area involved in facial recognition. Researchers at UCLA found that postmenopausal women on estrogen therapy were less depressed and angry and performed better on tests of verbal fluency, hearing, and working memory than did post-menopausal women who were not taking estrogen, and they outperformed men, too. By contrast, researchers at the University of Illinois found that women who had never taken HT had significantly more shrinkage in all brain areas than did women who took HT. They also found that the longer women took HT, the more gray matter, or brain cell volume, they had compared with women who weren't taking HT. These positive effects held and even increased the longer a woman took HT.

Each woman, of course, is an individual, and her brain is quite different not only from a man's but from other women's. This variation makes brain comparison studies between individuals difficult. One way around this difficulty, is to examine identical twins. A Swedish study looked at pairs of postmenopausal female twins, from age sixty-five to eighty-four, in which one twin took HT while the other did not over many years. The HT users had better scores on tests of verbal fluency and working memory than their twin sisters. The twins on HT, in fact, showed 40 percent less cognitive impairment, regardless of the type and timing of the hormone treatment.

Barbara Sherwin in Canada has also been studying the effects of estrogen on the brains of postmenopausal and post-hysterectomy women for over twenty-five years. In her research, estrogen treatment showed protective effects on verbal memory in healthy, forty-five-year-old, surgically menopausal women who had been given estrogen immediately after their operations. However, no effect was found when estrogen was given to older women years after their surgical menopause. These findings suggest that there is a critical time for initiating estrogen therapy following menopause. Sherwin believes these factors may explain why no protective effect of HT on cognitive aging was found in the WHIMS.

These recent studies on the brain-preserving effects of HT, and the contradictory results of the WHI and WHIMS, highlight some of the current controversies surrounding postmenopausal hormone therapy and the female brain.

FREQUENTLY ASKED QUESTIONS

What happens to my brain as I pass through menopause?

Menopause itself technically lasts for only twenty-four hours—

the day that is twelve months after your final period. The very next day you begin the postmenopause. The twelve months leading up to that single day of menopause make up the last months of the so-called perimenopause. At age forty to forty-five, the female brain begins the early phase of perimenopause, the two to nine years before the day of menopause. At this stage, the brain for some reason starts to become less sensitive to estrogen. The precisely timed dialogue between the ovaries and brain begins to get garbled. The biological clock controlling the menstrual cycle is wearing out. This difference in sensitivity causes the timing of the menstrual cycle to change, and periods start to come a day or two earlier. It can also cause menstrual blood flow to change. As the brain becomes less sensitive to estrogen, the ovaries may try to compensate some months by making even more estrogen, causing heavier menstrual flow. This decrease in sensitivity to estrogen in the brain can also trigger a cascade of symptoms that vary from month to month and year to year, ranging from hot flashes and joint pain to anxiety, depression, and changing levels of libido.

Depression is a surprisingly common problem in perimenopause. Researchers at the National Institute of Mental Health found that perimenopausal women have fourteen times the normal risk of depression. That risk is especially high during late perimenopause, the two years before menstruation stops. Why might this be so? At the maximum period of estrogen change, the neurochemicals and brain cells that are usually supported by estrogen—such as serotonin cells—have become disturbed. This perimenopausal depression can sometimes be treated with estrogen therapy alone if it is mild. Bottom line, the transition through perimenopause can be a time of vulnerability to mood instability and irritability because of the brain's changes in estrogen and stress sensitivity. Depression can come out of the blue, even for women who've never previously experienced it.

The lack of joy in life, in the absence of any real-life tragedy, may be caused by low estrogen in the brain, which in turn decreases neurochemicals such as the mood-elevating serotonin, norepinephrine, and dopamine. Irritability, lack of mental focus, and fatigue can be caused by low estrogen and made worse by lack of sleep. A major problem for many perimenopausal women is sleep—either with or without hot flashes. There's no time in your life when it's healthy to go without adequate sleep, but this is especially true when you're over age forty. Sleep is an essential renewing treatment for the brain. Unfortunately, erratic estrogen changes during perimenopause can disturb the female brain's sleep clock. If you don't sleep well for several days, it can be hard to concentrate; you may also become more impulsive and irritable than usual and say things you wish you hadn't. So this may actually be a good time to bite your tongue in order to protect relationships. All these symptoms of the perimenopause in my experience can usually be treated with a combination of estrogen, antidepressants, exercise, diet, sleep, and supportive or cognitive therapy.

Once a woman has officially passed through menopause, her brain has started readjusting to low estrogen. For most women, the disruptive symptoms of perimenopause now begin to abate, though a percentage of women, unfortunately, suffer for another five years or more. Fatigue, mood changes, interrupted sleep, "mental fog," and memory changes occur for some women, and more than about 15 percent still have hot flashes a decade or more past menopause. Some three out of ten postmenopausal women suffer from periods of low mood and depression, and up to eight out of ten experience fatigue. (All women with fatigue should have their thyroid checked.) Some studies, but not all, found that age-related cognitive functions, such as short-term memory, decline more quickly in the first five years after menopause.

In most cases, the female brain acclimates to lower levels of estrogen as the ovaries gradually retire. If a premenopausal woman has surgery to remove her uterus and ovaries, however, she'll plunge into menopause with no transition. The sudden loss of estrogen, as well as testosterone, can trigger symptoms including low energy, low self-esteem, and low libido, as well as severe mood and sleep changes along with hot flashes. Most women who have total hysterectomies can avoid these problems if they start on estrogen replacement therapy in the recovery room or even before surgery. Early treatment with estrogen can be especially important to protect memory function post-hysterectomy, as Barbara Sherwin's studies have suggested.

Should I take hormones for my brain, and what can I do to reduce my risk of stroke and breast cancer if I do?

Most doctors now feel that each woman should let her own symptoms at menopause or perimenopause be her guide. For many women, HT, especially with continuous estrogen, helps stabilize mood and improves mental focus and memory. Some women say estrogen therapy gives them back their sharp minds and makes them feel smart again. Other women report unpleasant side effects, such as menstrual bleeding, cramping, breast tenderness, and weight gain, which may cause them to discontinue the therapy.

So what's the best advice to date on HT? The Food and Drug Administration now recommends that women with menopause symptoms take the lowest dose of hormones for the shortest time possible, since scientists assume that lower doses are likely to be safer. The position statement by the Executive Committee of the International Menopause Society recommends that doctors not change their previous practices in prescribing hormone therapy to women at menopause or stop HT in any woman who is doing well on it because the WHI and WHIMS did not study women during the menopausal transition. Some American scientists, such as Fred Naftolin of Yale, are quite worried that doctors are now denying women the chance to take estrogen for prevention before it's too late. He says,

> So ... these menopausal symptoms are warnings of estrogen deficiency [that are] singing out to alert us of the need to test the idea of prevention by timely estrogen treatment. We must rethink the current American position on prevention of menopausal complications by estrogen and thereby afford women the [treatment and] scientific rigor that they deserve.

Some studies indicate that if you are more than six years past menopause you have lost your window for prevention and should not start HT. Bottom line, every woman needs to discuss her personal risks and benefits with a doctor who specializes in hormone therapies. Rogerio Lobo, an expert for thirty years in HT, states that "the appropriate use of hormones largely alleviates concerns about the increased risk of cardiovascular (CV) disease and breast cancer. The appropriate use of hormones pertains to treating younger, healthy women who have menopausal symptoms as well as using low-doses of hormones and switching to estrogen-only therapy whenever possible."

If you're suffering from symptoms that are disrupting your quality of life, you may want to consider a few years of hormones to ease your brain through this transition. It's not a moral issue; you're not a weak person if you happen to be in the large group of women who need some medical help to be their best selves during this hormonal transition.

And don't feel that you're making a decision today that will commit you to a particular treatment over the next forty years. You may want to continue HT after you get through the menopause transition, and you may not. Many new scientific discoveries and products regularly become available, and the race is on in the drug industry to develop estrogen-like drugs that help the brain and the bones without posing a risk to women's breasts, heart, uterus, and vascular system. There are also many nonhormonal and alternative medicines and treatments that can be very helpful—including exercise, SSRIs, soy, high-protein/low-calorie diet, vitamins E and B-complex, acupuncture, stress reduction, and meditation practice. The smart thing to do is keep informed and reevaluate your decision every twelve months.

If you do decide to take HT, be prepared for a period of trial and error. Responses vary greatly, so you'll have to test-drive different treatments in your own body. Some HT doctors like to start with bioidentical hormones, which are most like the ones your own ovaries produce. If for some reason these don't help you feel better, you should discuss other types of hormones; some women feel better on synthetic hormones or on patches, pills, gels, injectables, or pellets. If you still don't feel good or better, don't give up. Ask your doctor about alternatives or additions to hormones to treat your symptoms for the next year or two, including prescription serotonin drugs such as Effexor, Zoloft, or Prozac, herbal treatments, or exercise and relaxation therapies. The fact is, you know your own body best. Let your own symptoms be your guide. Above all, since new research is constantly emerging, plan to discuss whatever treatment you're currently using every year with your doctor—it's a good idea to set your appointment around your birthday so you won't forget.

One of the major reasons scientists believe women in the WHI and WHIMS who took HT had somewhat more stroke, dementia, and heart attacks was that taking estrogen on top of already clogged and aging blood vessels makes matters for the heart's and brain's blood vessels worse—especially since many of these women were smokers. If you decide to take hormone therapy, keep your blood pressure low, don't smoke, get at lest sixty minutes per week of increased-pulse cardiovascular exercise, keep your cholesterol low, eat as many vegetables as you can, take vitamins, decrease your stress, and increase your social support.

Weight gain, not brain functioning, is actually the biggest concern many women express about HT and the major reason they give worldwide for stopping the treatment. The hypothalamus controls our appetite. Since many of the changes during the menopause happen in this area of the brain, some scientists have speculated that the appetite-controlling cells are adversely affected by declining estrogen. To test whether weight gain was caused by HT, researchers in Norway studied ten thousand women ages forty-five to sixty-five who were and were not on hormone therapy. Their results showed that weight gain is not linked to HT. Instead they found that changes in a woman's diet and physical activity, both of which may have to do with changes in her hypothalamus at menopause, are the cause of weight gain.

A NOTE ABOUT HORMONE THERAPY: ESTROGEN WITH OR WITHOUT PROGESTERONE

Estrogen-only therapy, without progesterone, it's important to note, is appropriate only for postmenopausal women who've had hysterectomies. It's not the same as hormone replacement therapy (HT), with progesterone, which is prescribed for women who still have a uterus. There's an important difference: HT with progesterone keeps estrogen from building up the uterine lining and possibly producing cancer cells. Progesterone can be taken in pill form combined with estrogen or as an intrauterine device with progesterone or vaginal gel. Progesterone, however, seems to counter some of the positive effects of estrogen in the female brain. Just as progesterone reverses the growth of unwanted cells in the uterus, it seems to reverse some of the growth of new connections in the brain. As a result, the brain benefits of HT with progesterone are a matter of controversy. If a woman is able to take estrogen alone because she has no uterus, she can get all the benefits of estrogen she had at the best part of her menstrual cycle—all the time, but without the PMS-causing progesterone. Some women who do not tolerate progesterone but still have a uterus can have annual removals of their uterine lining through a procedure called dilation and curettage (D & C) or endometrial ablation. They can also get annual vaginal ultrasounds of the uterine lining to make sure it isn't growing. Women taking the lowest doses of estrogen HT do not usually need to take progesterone even if they still have a uterus.

It's not until many years after menopause that the natural processes of aging start having a noticeable effect on the functioning of the female brain. Some memory decline does start as early as age fifty, but it is usually not bothersome. Hormone therapy may or may not help slow it down. Many of these aging processes involve decreased blood supply and a breakdown in the body's ability to repair damage.

It is now clear that estrogen keeps blood vessels in the brain healthy. Researchers at the University of California, Irvine, found that estrogen did this by increasing the efficiency of the mitochondria in the brain's blood vessels, perhaps explaining why premenopausal women have lower rates of stroke than men their age. Research at Children's Hospital in Pittsburgh, Pennsylvania, also discovered a sex difference in the way brain cells die after injury. Levels of glutathione, a molecule that helps brain cells survive oxygen deprivation, remain stable in females after a brain injury, but they drop up to 80 percent in males, resulting in greater brain cell death. It may be that male and female brain cells die in different ways following established sexspecific biological patterns and pathways that may be related to why women live longer than men.

Sex differences appear in other aging processes, too. Estrogen and progesterone, for example, seem to help repair and maintain the connecting cables between brain areas. As our brains age and our bodies stop repairing these connections, we lose white matter, and our brains process and send information more slowly or not at all. As a result, some signals get weaker, changing the pathways, patterns, and speed in our aging brains.

One process that often slows down noticeably is memory retrieval. This is common in the older brain even though no specific disease or dementia is present. Alzheimer's is one of a group of dementia diseases that gradually destroy brain cells and impair mental function. Alzheimer's makes sticky plaques in the brain, decreasing the ability of brain cells to communicate with one another and eventually killing them. Although men tend to be more vulnerable to age-related memory loss than women, postmenopausal women, it turns out, have three times more risk than men for developing Alzheimer's disease. Scientists don't yet understand this gender difference but suspect it may have to do with older men's brains having more testosterone and estrogen than those of postmenopausal women who don't take HT. Careful studies of the brains in an animal model of Alzheimer's have shown deficient levels of estrogen. It remains a mystery, nevertheless, why women are more susceptible to this disease even after correcting for the fact that on average they live longer.

Studies indicate that starting estrogen replacement therapy early in menopause, when neurons are healthy, reduces the risk of Alzheimer's disease. However, estrogen therapy initiated once the disease has developed or decades after menopause offers no benefit. Evidence from animal experiments and human studies also suggests that estrogen therapy may be able to delay dementia symptoms and brain aging in females. The idea that estrogen therapy may help prevent some cases of Alzheimer's in women is an attractive one but remains to be proven.

For women—even those past menopause—staying socially connected and supported is an important way to reduce the stresses of living alone and getting older. Women respond to stress differently than do men and get more benefit from social support.

Many activities can counter the effects of aging on the brain. Researchers at Johns Hopkins University found that women and men over age sixty-five who had the widest variety of activities had the lowest rates of dementia. Physical exercise, such as walking and bike riding, helped, but so did mental exercises, such as playing cards. As our bodies age, it's important to stay active on many levels, and it's diversity, not intensity, that may be key.

COPING WITH ANOTHER BRAIN DRAIN: TESTOSTERONE LOSS

Unfortunately, estrogen loss isn't the only brain drain for females around menopause. By age fifty, many women have lost up to 70 percent of their testosterone. This is because not only do the ovaries stop manufacturing as much at menopause but the adrenal glands, which provide 70 percent of a woman's androgens and testosterone, made as a prehormone called DHEA, during her fertile years, have greatly decreased their production, too, resulting in a hormone transition called "adrenopause." After menopause, the adrenal glands—even with their diminished production—supply over 90 percent of a woman's androgens and testosterone. Both men and women, in fact, go through this testosterone and androgen loss from the adrenal gland, as some of the adrenal cells die starting at about age forty. By age fifty, men have lost half of their adrenal testosterone and 60 percent of the testosterone produced by the testes when they were young. Men's sex drive, as a result, often declines in these years. Since testosterone is required to stimulate sexual interest in the brain, the plunge in testosterone after menopause can cause women to feel little or no interest in sex.

Males, for most of their adulthood, produce ten to one hundred times as much testosterone as females do. Their testosterone levels range from 300 to 1000 (picograms per milliliter), compared with 20 to 70 for women. Even though men's testosterone drops 3 percent a year on average from its high, at age twenty-five, it usually stays well above 350 into middle age and beyond—and 300 picograms per milliliter is all men need to maintain sexual interest. It takes much less testosterone to spark sexual urges in a woman, but she does need enough to trigger her brain's sex center. Women's youthful testosterone high is at age nineteen, and by the age of forty-five or fifty, women's levels have dropped by up to 70 percent— leaving many with very low testosterone levels. In these cases, like a car that's out of gas, the sex center in the hypothalamus doesn't have the chemical fuel it needs to ignite sexual desire and genital sensitivity. The physical and mental engines of sexual arousal stall.

Complaints about women's sexual interest and performance are extremely common at all ages. Four in ten American women—nearly half—are unhappy with some aspects of their sexual lives, and between the ages of forty and fifty, that number climbs to six in ten. Some of the most widespread complaints in women during and after the perimenopause are diminished sex interest and arousal, difficulty achieving orgasms, weaker orgasms, and aversion to physical or sexual touch. Millions of women suddenly see their sex drive disappear— and researchers have found strikingly similar patterns all over the world. The biological reasons for this decline are profound hormonal changes in the brain. The estrogen, progesterone, and testosterone surges from the ovaries that formerly marinated the brain are now ending. Androgen and testosterone production by the adrenal glands and ovaries, which surged around puberty and remained high into a woman's twenties and early thirties, dwindles by about 2 percent per year, until by the age of seventy or eighty we have only 5 percent of what we had when we were twenty. Libido in women decreases with age starting in the third decade of life and is especially prevalent if women have had their ovaries removed.

Sexual intercourse and interest in sex in women begin to decline in the fourth and fifth decades. Most women who have sexual partners at menopause continue to have sex. Studies in nursing homes have shown that a quarter of women ages seventy to ninety still masturbate. For those who have experienced declining sexual interest and want to dial it back up again, restoring testosterone to more youthful levels with gels, creams, or pills may help. Until recently, however, medical science paid scant attention to testosterone deficiency in females. Doctors feared, instead, that women might have too much of this chemical traditionally associated with masculinity and develop unnatural male traits, such as facial hair, aggression, and deep voices. In large measure because of this bias, there's been almost no focus until recent years on the real and troubling effects for women of too little testosterone.

WHAT TO DO ABOUT SEXUAL COMPLAINTS AND HOW TO GET HELP

Those who grew up in the culture of the feminist and sexual revolutions and beyond believe that hot, passionate, satisfyingly orgasmic sex is something to which women should feel entitled. Over the last two or three decades, the stereotype of the easily aroused, enthusiastically sexual, even predatory female has replaced the more traditional view of the demure woman who has to be seduced or loosened up with alcohol. But this new woman is a fiction in much the same way her reticent forerunner was. Unfortunately, the truth is, many women discover at the beginning of menopause that good sex is not only hard to find but also physically challenging, impossible, or unappealing. We may suddenly find ourselves grappling with low or no sex drive, arousal problems, or the inability to have an orgasm—physical changes that can be surprising and discouraging, to say the least. I see women with these issues every day in my clinic. My patients complain that it has been hard for them to find a doctor who is knowledgeable about the female sexual response—how it can vary with hormones and from person to person, and how it can change dramatically over the course of a woman's life. To this day, most medical schools don't teach a required course in female sexual response.

Even gynecologists, who specialize in body parts below the waist, have few answers for women with sexual problems and often find no physical reasons for their symptoms. As a result, they tend to dismiss these issues as "simply part of getting older"—ignoring the toll they can take on women's relationships and quality of life. Psychiatrists and couples' therapists can be equally ill-equipped to offer help. They tend to see the problem as all in the head—the result of stress in the relationship or long-term problems with intimacy. A classic response to these issues has been psychoanalysis—putting a woman on the couch for seven to ten years to get to the roots of her unnatural "frigidity" or psychological "resistance" to sex. This approach is mostly wrong-headed because the reason for these feelings at this stage of life is not a psychological conflict; it is a normal biological and psychological response to hormonal changes.

One key to restoring female libido is testosterone replacement therapy. Researchers discovered its effectiveness decades ago, but medical science in the United States has largely ignored or forgotten this information. Forty years ago, in the 1970s, doctors at the University of Chicago experimentally gave large amounts of testosterone to female patients suffering from breast cancer. Their thinking was that the hormone would lower the women's levels of estrogen, which can promote cancer. It didn't, but the subjects experienced a tremendous increase in their libidos and orgasmic capacity. The same effect was seen in the 1980s by Barbara Sherwin of McGill University. Sherwin replaced testosterone in women who had their ovaries removed. Those who didn't get the hormone reported steep declines in their libidos; those who did get treatment reported that their sexual interest soon returned.

Studies are finally beginning to look at therapies for sexual dysfunction in women above the groin, targeting the female brain centers that are linked to pleasure and desire. And the treatment that does work—testosterone replacement—is finally moving into acceptance. In recent years, testosterone supplements have been a wildly popular treatment regimen for men. Only very recently, however, have doctors begun dispensing testosterone gels, patches, and creams for women patients. I've been prescribing testosterone replacement for women since 1994, and the results have been mostly positive.

When women complain of low libido, testosterone replacement therapy often brings their sexual interest back to par. We know that by giving testosterone we can increase a woman's urge to masturbate and shorten her time to orgasm, but not necessarily increase her desire for partner sex. For some women, testosterone can improve sexual interest dramatically, but the hormone may not be the panacea we once thought for improving sexual interest in all women. Even men are discovering that testosterone or Viagra is not the magic bullet promised by the drug companies. However, there is no question that having a barely measurable or zero level of testosterone in men or women can be a cause of sexual dysfunction. This condition can be treated in both sexes with testosterone therapy. Women who complain of a lack of sexual interest—whether they are pre-menopausal or postmenopausal—deserve a trial of testosterone just as most doctors would prescribe for a man.

In addition to its effects on the brain's sexual center, testos-terone promotes mental acuity as well as muscle and bone growth. On the downside, it can contribute to thinning hair, acne, body odor, facial hair growth, and a lower voice. But the effects of testosterone on the brain—increased mental focus, better mood, more energy and sexual interest—are the reasons many men and women who take it say they are willing to assume the downside risks.

The Female Brain and Postpartum Depression

O NE OUT OF TEN female brains will become depressed within the first year after giving birth. For some reason, this 10 percent of women have brains that do not entirely rebalance themselves after the massive hormone changes that follow giving birth. Postpartum psychiatric changes can range from maternity blues to psychosis, but the most common is postpartum depression. Women suffering from this condition are thought to have an increased genetic susceptibility to becoming depressed as a result of hormone changes. Ken Kendler of Virginia Commonwealth University found that there may be genes that alter the risk for depression in a woman's response to cyclic sex hormones, particularly in the postpartum period. Such genes would affect women's risk for major depression but would not be active in men because men lack the relevant hormonal changes. These results suggest a role for changes in estrogen and progesterone in precipitating mood symptoms among women with postpartum depression.

These 10 percent of women seem to get depressed postpartum for multiple reasons. The brain has had its stress-response "brakes" on during pregnancy; suddenly, after birth, they come off again. For 90 percent of women, the brain can return to a normal stress response, but for vulnerable women it is unable to do this. A vulnerable woman's brain ends up hyperreactive to stress and she makes too much of the stress hormone cortisol. Her startle reflex will be up, she'll be jumpy, small things will seem like enormous problems. She'll be hypervigilant over the baby, hyperactive, and unable to get back to sleep after feeding the baby at night. She'll be walking around day and night jittery, as though her finger is in a light socket even though she is exhausted.

The well-known predictors of depression after giving birth include a previous depression, depression during pregnancy, lack of proper emotional support, and high stress in the home. Women with postpartum depression were also struggling with their identities in the face of their new roles as mothers. They express feelings of loss of a sense of who they are as individuals. They feel overwhelmed by the responsibility for their child. They are coping with feeling abandoned by their partners and others close to them who aren't supporting them enough, unreasonable worries that their child will die, and breast-feeding problems. They feel like "bad mothers," but they never blame their child. Most mothers are reluctant to speak about their feelings and assign their moods to personal weakness rather than illness. They are struggling to keep their equality with their partners and to get the fathers involved in child care.

The transition to parenthood is often accompanied by depression and stress. It's a whole new life and reality, so feeling rocked by the experience is understandable. In addition, mothers' drastic hormone changes have created quantum shifts in their reality several times in less than a year. Women who are vulnerable to depression and stress may have a harder time rebalancing from these shifts. And if you're having trouble rebalancing, a fussy child and no sleep will only increase your vulnerability to depression. For some women, these feelings of stress don't peak until up to twelve months postpartum.

Furthermore, postpartum depressive symptoms often remain hidden. Women are ashamed because they are expected to be so happy at the birth of their child. So it is important to understand the complexity of postpartum depressive mood as struggling with rebalancing brain hormones, a new identity, breast feeding, sleep, the child, and the partner.

Some scientists feel that breast feeding may be protective against postpartum depression in certain women. During lactation, mothers exhibit lower neuroendocrine and behavioral responses to several types of stressors, except possibly those representing a threat to the infant. This ability to filter relevant from irrelevant stimuli is viewed as adaptive for the mother-infant dyad, and the inability to filter stressful stimuli can be associated with the development of postpartum depression.

The good news is that treatment is available, and it is effective. The brain chemicals such as serotonin that help support mood and well-being are running low after giving birth and the postpartum brains in depressed moms have a deficit. Medications and hormones can help return their brains to normal. A consensus among experts in postpartum depression recommends—for women with severe symptoms—antidepressant medication combined with other treatment modalities, such as supportive talk therapy.

The Female Brain and Sexual Orientation

HOW DOES SEXUAL orientation get wired in the female brain? There are many variations in the female brain that lead to individual skill sets and behaviors. Our genetic variations and the hormones present in our brains during fetal development are the cornerstone of the female brain. Life experiences then play upon our particular female brain circuits to reenforce individual differences. One variation that appears on a continuum in females is same-sex romantic attraction. This is estimated to occur in five to ten percent of the female population.

The female brain is only half as likely to be wired for same-sex attraction as is the male brain. Therefore men are twice as likely as women to be gay. Biologically, genetic variations and hormonal exposure in both male and female brains are thought to lead to same-sex attraction, but the origins in women appear to be different than in men. Most brain studies have been done on the difference between gay and straight males, and only recently have studies in females begun to emerge. Sexual orientation in females occurs along more of a continuum than in males, with females reporting more bisexual interests. Psychosocial studies have also shown that gay women have higher self-esteem and quality of life than do gay men. This may be because it is socially easier to be a gay woman than a gay man.

Sexual orientation does not appear to be a matter of conscious self-labeling but a matter of brain wiring. Several family and twin studies provide clear evidence for a genetic component to both male and female sexual orientation. We know that prenatal exposure to an opposite-sex hormonal environment, like testosterone in a genetically female brain, leads the nervous system and brain circuits to develop along more male-typical lines. This prenatal hormonal environment has enduring effects on behavioral traits like rough-and-tumble play and sexual attraction.

Core gender identity and sexual orientation were assessed in one study along with recalled childhood gender role behavior in women who were exposed to higher levels of testosterone in utero. They recalled more male-typical play behavior as children than did women not exposed to fetal testosterone. These women also reported more same-sex attraction and were more likely to be gay or bisexual.

One study examined brain-wiring differences, as indicated by the "startle response" between gay versus straight women. They found that gay women had a lower startle response—in a similar range to most men—indicating brain-wiring differences between straight and gay women. Gay versus straight women showed a less sensitive auditory response—a male-typical pattern. Female brains usually perform better than male brains in tests of verbal fluency. Gay women showed opposite-sex shifts in their verbal fluency scores—scoring in a range intermediate between males and females. Gay women identified as "butch" versus "femme" showed a range of scores intermediate between males and females. And straight women scored better overall in verbal fluency tests than their gay female counterparts. This indicates that these differences in brain circuitry are on a continuum in the female brain. These scientific findings indicate that the wiring of the female brain for sexual orientation occurs during fetal development, following the blueprint of that individual's genes and sex hormones. The behavioral expression of her brain wiring will then be influenced and shaped by environment and culture.

NOTES

The chapter notes are the result of many years of research, thought, and synthesis of ideas. I have gathered the work of many scientists in various disciplines in order to arrive at this understanding of the female brain. As a result, there are notes containing multiple references, reflecting the various sources I used to arrive at the theory expressed in the text.

Here in the notes where there are multiple authors of a paper or book, I have listed only the first author and the year of publication. Where more than one reference is used, I have arranged them in chronological order. The full citation may be found in the References section.

INTRODUCTION: What Makes Us Women

23 "... *perception, thoughts, feelings and emotions.*": Nishida 2005; Orzhekhovskaia 2005; Prkachin 2004; see Chapter 6, "Emotions."

24 "... *in women compared with men.*": Blehar 2003; Madden 2000; Weissman 1993.

25 "... *pre-menstrual brain syndrome.*": Schmidt 1998; see Chapter 2, "Teen Girl Brain."

25 "... *twenty-five percent every month.*": Woolley 1996, 2002.

25 "... *or bite someone's head off.*": see Chapter 2, "Teen Girl Brain."

27 "... *brain sensitivities to stress and conflict.*": Shors 2001.

27 "... *activated to complete the task.*": Bell 2006; Jordan 2002.

28 "... *by using different brain circuits.*": Tranel 2005; Jordan 2002.

28 "... *percent more neurons than men.*": Witelson 1995; see also: Knaus 2006; Plante 2006; Wager 2003.

28 "... *language and observing emotions in others.*": Goldstein 2005; Giedd 1996.

28 "... *four times on her hottest days.*": See Chapters 4, "Sex," and 7, "The Mature Female Brain."

28 "... *as just two people talking.*": See Chapter 3, "Love and Trust."

28 "... *triggers aggression—the amygdala.*": Cahill 2005; Giedd 1996; Witelson 1995.

28 "... *try anything to defuse conflict.*": Campbell 2005; see Chapter 6, "Emotions."

29 "... *stress experienced in the ancient wild.*": See Chapters 2, "Teen Girl Brain," 3, "Love and Trust," and 5, "The Mommy Brain."

29 "... *endangered by impending catastrophe.*": see Chapters 2, "Teen Girl Brain," 3, "Love and Trust," and 5, "The Mommy Brain."

30 *". . . such as math and science.":* Blinkhorn 2005; Cherney 2005; Haier 2005; Jausovec 2005.

30 *". . . are not plausibly, culturally determined.":* Summers 2005

30 *". . . scientific capacity is nonexistent.":* Spelke 2005.

31 *". . . or hours of computer time.":* see Chapter 2, "Teen Girl Brain."

31 *". . . careers because of lack of aptitude.":* Lawrence 2003, 2006; Babcock 2004.

31 *". . . being able to defuse conflict.":* See Chapter 6, "Emotions."

ONE: THE BIRTH OF THE FEMALE BRAIN

35 *". . . hard-wired into the animal brain.":* Hines 2002.

36 *". . . both genes and sex hormones.":* Arnold 2004.

36 *". . . the sex and aggression centers.":* Sur 2005.

36 *". . . areas that process emotion.":* See Chapter 6, "Emotions."

36 *". . . communication than he will.":* Tannen 1990.

37 *". . . baby to do is study faces.":* McClure 2000.

37 *". . . terribly wrong with my son.":* Bowlby 1980.

38 *". . . will not increase at all.":* Leeb 2004.

38 *". . . had too many botox injections.":* Herrera 2004.

39 *". . . symbiosis with their mothers.":* Silverman 2003.

39 *". . . faster, by one to two years.":* Nosip 2004.

39 *". . . social approval of others very early.":* Camras 1990.

40 *". . . they are forbidden to touch.":* Rosen 1992.

40 *". . . and hearing human vocal tones.":* Plante 2005; Bachevalier 1990, 1991.

40 *". . . in the human voice than boys.":* Plante 2005.

41 *". . . a baby girl's behavior.":* Baron-Cohen 2005.

41 *". . . their mothers than do boys.":* Weinberg 1999.

41 *". . . face—than male newborns do.":* McClure 2000.

41 *". . . those who look sad or hurt.":* McClure 2000.

42 *". . . twenty-four months long for girls.":* Grumbach 2005, personal communication; Soldin 2005; Bachevalier 1989, 1991.

42 *". . . brain for reproductive purposes.":* Grumbach 2005.

42 *". . . communication, even tending and caring.":* Yamamoto 2004.

42 *". . . mother's nervous system into her own.":* Leckman 2004.

43 *". . . a fearful, anxious one."*: See Chapter 5, "The Mommy Brain."

43 *". . . nurturing their mothers are."*: Cameron 2005; Cooke 2005; De Kloet 2005; Fish 2004; Zimmerberg 2004; Kinnunen 2003; Champagne 2001; Meaney 2001; Francis 1999.

43 *". . . females and non-human primates."*: Kajantie 2006; Capitanio 2005; Kaiser 2005; Gutteling 2005; Wallen 2005; Huot 2004; Lederman 2004; Ward 2004; Morley-Fletcher 2003.

44 *". . . micro-circuitry at the neurological level."*: Leckman 2004.

44 *". . . their mother's nervous system."*: Cameron 2005; Roussel 2005; Ward 2004.

44 *". . . were measured in goat kids."*: Roussel 2005.

45 *". . . important in the twenty-first century."*: Campbell 2005.

45 *". . . in preserving harmonious relationships."*: Sanchez-Martin 2000.

46 *". . . response instead of forging ahead."*: Tannen 1990.

46 *". . . conflict, or displays of status."*: Campbell 2005; Tannen 1990.

46 *". . . they typically don't use it."*: Tannen 1990.

47 *". . . is part of their makeup."*: Maccoby 1998.

47 *". . . or commands given by girls."*: Maccoby 1998.

47 *". . . more easily socially handicapped."*: Baron-Cohen 2005; Campbell 2005.

47 *". . . emotional and social sensitivity."*: Baron-Cohen 2005; Herba 2004.

47 *". . . more estrogen than boys."*: Grumbach 2005 personal communication.

47 *". . . every culture that's been studied."*: Maccoby 1987.

48 *". . . brain differences may be one reason."*: McClure 2000.

49 *". . . they'll just stop playing."*: Maccoby 1998.

49 *". . . involve any high-spirited boys."*: Maccoby 1998, 2005 personal communication; Fagot 1985; Jacklin 1978.

49 *". . . nurturing or caregiving relationships."*: Maccoby 1998.

49 *". . . of territory and physical strength."*: Maccoby 1998.

49 *". . . quality of their social relationships."*: Knickmeyer 2005.

49 *". . . their usual interest in infants."*: Wallen 2005.

50 *". . . play than do average females."*: Wallen 1997, 2005; Goy 1988.

50 *". . . males than to those of females."*: Pasterski 2005; Berenbaum 1999.

51 *". . . dressing up in princess costumes."*: Hines 1994, 2004.

51 *"... traits that are typically female.":* Hines 2003; Berenbaum 1999, 2001.

51 *"... testosterone that gets into the fetal brain.":* Arnold 2004.

52 *"... and more neurons to that activity.":* McClure 2000; Fivush 1989.

52 *"... and less adventurous than boys.":* Golomboch 1994.

53 *"... people and with our environment.":* Cameron 2005; Iervolino 2005.

53 *"... susceptibility to environmental influences.":* Knafo 2005

53 *"... child development is inextricably both.":* ibid.

54 *"... boys—but don't be fooled.":* Leppänen 2001.

54 *"... both sexes have brain circuits for it.":* Campbell 2005.

54 *"... reflecting their unique brain circuitry.":* ibid; Archer 2005; Ferguson 2000; Crick 1996.

55 *"... stereotype born out of the contrast with boys.":* Knight 2002; Davidson 1996; ibid Archer 2005.

55 *"... the prison system will confirm.":* Campbell 2005.

56 *"... to feel safe and protected.":* ibid; Dodge 1982.

TWO: TEEN GIRL BRAIN

57 *"... and obsesses over her looks.":* Giedd 1996, 2004, 2005 personal communication.

58 *"... for independence and identity.":* Nelson 2005; Schweinsburg 2005; Romeo 2001, 2002.

58 *"... nurturing those around them.":* McClure 2000.

59 *"... well-being during these rocky years.":* Udry 2004; Barmeister 2000.

60 *"... monthly waves from her ovaries.":* Speroff 2005.

60 *"... that difference is even greater.":* Sokhi 2005; Open-Speech-Recognizer 2005.

60 *"... new estrogen and progesterone fuel.":* Goldstein 2005; Giedd 1997.

61 *"... puberty and into early adulthood.":* Schweinsburg 2005; Luna 2004.

61 *"... until she passes through menopause.":* Jasnow 2006; Hodes 2005; Shors 2005.

61 "... *their daily levels of cortisol.*": Morgan 2004; Stroud 2004.

61 "... *once they have entered puberty.*": Stroud 2004.

61 "... *stress responsivity in the hippocampus of females.*": Taylor 2006; Young 2006 personal communication; Viau 2004, 2005, 2006 personal communication; Agrati 2005; Putnam 2005; Shors 2001.

61 "... *creation of protective social networks.*": Taylor 2000, 2006; Kudielka 2005; Klein 2002; Stroud 2002; Bebbington 1996.

61 "... *she hates relationship conflict.*": Kiecolt-Glaser 1996, 1998.

61 "... *triggered by social rejection.*": Stroud 2002

61 "... *social stress on a weekly basis.*": Morgan 2004; Kirschbaum 1999; Kudielka 1999.

61 "... *react with increased irritability.*": Kudielka 2004, 2005.

62 "... *to the stress hormone cortisol.*": Stephen 2006; Cooke 2005; Mowlavi 2005; Morgan 2004; Rose 2004; Roca 2003; Berkley 2002; Young 1995, 2002; Cyranowski 2000; Kirschbaum 1999; Altemus 1997; Keller-Wood 1988.

62 "... *for talking, flirting, and socializing.*": Matthews 2005; Salonia 2005; Uvnäs-Moberg 2005; Cameron 2004; Ferguson 2001; Giedd 1999; Paus 1999; Turner 1999; Gangestad 1998; De Wied 1997; Slob 1996; Alexander 1990; Cohen 1987.

63 "... *vocabularies than do boys.*": Hyde 1988.

63 "... *in a social setting.*": Tannen 1990.

63 "... *weeks without vocalizing at all.*": Wallen 2005.

63 "... *and downs and stresses of life.*": Rose 2006; Maccoby 1998; Dunbar 1996.

64 "... *pleasure centers in a girl's brain.*": Glazer 1992.

64 "... *increases dopamine production in girls.*": Forger 2004, 2006; Dluzen 2005; Walker 2000.

64 "... *is triggered by intimacy.*": Uvnäs-Moberg 2005; Turner 1999; Whitcher 1979.

64 "... *reinforcement for social bonding.*": Depue 2005; Johns 2004; Jones 2004; Motzer 2004; Heinrichs 2003; Martel 1993.

64 "... *her urge for intimacy is also peaking.*": Goldstein 2005; Uvnäs-Moberg 2005.

65 "... *than they did before puberty.*": Dunbar 1996

66 "... *testosterone marinate the boys' brains.*": Bradley 2005.

66 *". . . involves sports or sexual pursuit."*: Pennebaker 2004; Rowe 2004; Sanchez-Martin 2000.

67 *". . . especially true in the teenage female brain."*: Jasnow 2006; Bertolino 2005; Hamann 2005; Huber 2005; Pezawas 2005; Sabatinelli 2005; Viau 2005; Wilson 2005; Phelps 2004.

68 *". . . rejection than does the male brain."*: Ochsner 2004; Levesque 2003; Zubieta 2003.

68 *". . . a positive boost from it."*: Maccoby 1998.

68 *". . . her friend will be their last."*: Kiecolt-Glaser 1996, 1998.

68 *". . . hormone cortisol takes over."*: Kudielka 2005; Stroud 2002, 2004; Klein 2002; Bebbington 1996.

69 *". . . intimate relationships with others."*: Mackie 2000; Josephs 1992.

69 *". . . formation of cliques and clubs."*: Jasnow 2006; Rose 2006.

69 *". . . described by W. B. Cannon in 1932."*: Cannon 1932.

69 *". . . male response to threat and stress."*: Taylor 2006, 2000.

69 *". . . the demands of an imminent threat."*: Sapolsky 1986, 2000.

70 *". . . brain areas for physical action."*: Campbell 2005; O'Connor 2004; Collaer 1995; Olweus 1988; Hyde 1984.

70 *". . . once they've formed maternal bonds."*: Keverne 1999; Mendoza 1999.

70 *". . . networks that may aid in this process."*: Taylor 2000.

70 *". . . him away with threatening cries."*: Dunbar 1996.

71 *". . . maternal behavior for younger females."*: Silk 2000; Wrangham 1980.

71 *". . . success at passing on their genes."*: Silk 2003.

72 *". . . clock cells, the suprachiasmatic nucleus."*: Toussan 2004.

72 *". . . brain cells that control breathing."*: Behan 2005.

72 *". . . and more sleeping time overall."*: Roenneberg 2004.

72 *". . . of all their brain circuits."*: Campbell 2005.

72 *". . . that will last until after menopause."*: Roenneberg 2004.

73 *". . . think more quickly and more agilely."*: Monnet 2006; Routtenberg 2005; Uysal 2005.

74 *". . . progesterone in the last two weeks."*: Kuhlmann 2005; Routtenberg 2005; Sa 2005; Cameron 1997, 2004; Weissman 2002; Woolley 1996.

74 *". . . momentarily upset, stressed, and irritable."*: Kajantie 2006;

Goldstein 2005; Protopopescu 2005; Kirschbaum 1999; Tersman 1991.

74 *"... right before their periods begin."*: Birzniece 2006; Kuhlmann 2005; Rubinow 1995.

75 *"... week two—of their cycles."*: Birzniece 2006; Sherwin 1994; Phillips 1992.

76 *"... spectrum of discomfort as a seizure."*: Smith 2004

76 *"... before the onset of bleeding."*: Altemus 2006; Mellon 2004, 2006; Schmidt 1998.

76 *"... maximal hormone withdrawal."*: Jovanovic 2004; Toufexis 2004.

76 *"... extremely edgy and easily upset."*: Parry 2002.

76 *"... have fewer serotonin brain cells."*: Bethea 2005; Zhang 2005; Cameron 2000; Williams 1997.

76 *"... crying and rage can plague them."*: Bennett 2005; Lu 2002; Cyranowski 2000; Young 1995. (Drugs such as Prozac, Zoloft, and other antidepressants lift the brain's mood chemicals, including serotonin.)

77 *"... two weeks of each month."*: Korol 2004.48 *"... and from week to week."*: Goldstein 2005; Protopopescu 2005; Bowman 2002.

77 *"... premenstrual dysphoric disorder (PMDD)."*: Klatzkin 2006.

78 *"... calming right before the period starts."*: Smith 2004; Silberstein 2000.

78 *"... and progesterone during the cycle."*: Roca 1998, 2003; Schmidt 1998.

78 *"... removing the ovaries surgically."*: Parry 2002.

78 *"... that were upsetting her brain."*: Joffe 2006 personal communication; Kirschbaum 1999.

78 *"... her serotonin levels stabilized."*: Kurshan 2006; Griffin 1999; Kirschbaum 1999; Tuiten 1995.

78 *"... mood and sense of well-being."*: Freeman 2004; Luisi 2003.

79 *"... highly responsive at puberty."*: Toufexis 2004.

79 *"... thin and immature."*: Giedd 2005 personal communication; Cardinal 2004; Lewis 2004.

79 *"... often becomes overwhelmed."*: see Chapter 6, "Emotions."

80 *"... function reliably under stress."*: Giedd 2005.

82 *"... impulsive time for many girls."*: Young 2004.

82 "... *prefrontal cortex may function normally.*": Roca 1998, 2003; Altemus 2001.

82 "... *event feel like a catastrophe.*": Arnsten 2004; Berenbaum 2004.

82 "... *amygdala can prove difficult.*": Arnsten 2004.

82 "... *when they're under stress.*": Genazzani 2005; Dobson 2003.

83 "... *ratio for depression doubles.*": Staley 2006; Weissman 1993, 2000, 2005; Blehar 2003; Mazure 2003; Maciejewski 2001; Kendler 2000.

83 "... *to suffer from depression.*": Weissman 1999, 2002; Hayward 2002; Born 2002.

83 "... *role in female depression.*": Muller 2002.

84 "... *higher risk for clinical depression.*": Zubenko 2002.

84 "... *pivotal—among teenage girls.*": Archer 2005; Fry 1992; Burbank 1987.

84 "... *that among teen boys.*": Campbell 2005, 1995.

84 "... *rumors to undermine a rival.*": Holmstrom 1992; Eagly 1986.

85 "... *and females are androgens.*": Carter 2003.

85 "... *and twenty-one in males.*": Vermeulen 1995.

85 "... *have sexual intercourse earlier.*": Netherton 2004; Halpern 1997.

85 "... *over boys and other girls.*": Dreher 2005; Pinna 2005; Weiner 2004; Bond 2001; Udry 1977.

85 "... *levels in women and teens.*": Underwood 2003.

86 "... *estrogen, testosterone, and androstenedione.*": Cashdan 1995, Schultheiss 2003.

THREE: LOVE AND TRUST

89 "... *brain's love-drive by evolution.*": Rhodes 2005, 2006; Brown 2005.

89 "... *hormone that stokes sexual desire.*": Fisher 2005; see Chapter 4, "Sex."

89 "... *try to hook up with her.*": Emanuele 2006.

90 "... *help our offspring survive.*": Buss 1993.

90 "... *as neurological love circuits.*": Esch 2005.

91 "... *engineering of the human mind.*": Fisher 2005; Aron 2005.

92 "... *evolutionary psychologist David Buss.*": Buss 1990.

93 "... *a savvy investment strategy.*": Trivers 1972.

93 "... *triples children's survival rate.*": Hill 1988.

93 "... *shelter and other resources.*": Carter 2004; Reno, 2003.

93 "... *loved and worshiped her back.*": Botwin 1997.

94 "... *curvy, hourglass figures.*": Schutzwohl 2006; Singh 1993, 2002.

94 "... *females to have sex with.*": Schmitt 1996.

94 "... *visual clues to their fertility.*": Singh 2002.

95 "... *narrower than their hips.*": ibid.

95 "... *in size to their hips.*": Singh 1993.

95 "... *pregnancy radically alters her silhouette.*": ibid.

96 "... *still turned on full force.*": Carter 1998; see Chapter 6, "Emotions."

96 "... *when looking for a mate.*": Haselton 2005.

96 "... *trustworthy than they really are.*": Buss 1995; Tooke 1991.

96 "... *to agree to have sex.*": Haselton 2005.

96 "... *play earlier than boys.*": Maccoby 1998.

96 "... *eye gaze, and facial expressions.*": see Chapter 6, "Emotions."

96 "... *sexual behavior as is the male.*": Carter 1997; Kanin 1970.

96 "... *and months of a relationship.*": Hrdy 1997.

96 "... *high-level visual processing areas.*": Aron 2005; Brown 2005; Brown 2005 personal communication; see Chapter 6, "Emotions."

96 "... *'at first sight' more easily than women.*": Aron 2005; Fisher 2005 personal communication; Fisher 2004.

97 "... *on the test of passionate love.*": Fisher 2005.

98 "... *intoxication, thirst, and hunger.*": Aron 2005; Small 2001; Denton 1999.

98 "... *estrogen, oxytocin, and testosterone.*": Aron 2005.

98 "... *craving the next fix.*": Insel 2004.

98 "... *are running full blast.*": Pittman 2005; Debiec 2005; Huber 2005; Kirsch 2005; Bartels 2004.

98 "... *become addicted to love.*": Insel 2003.

99 "... *tendency to trust the hugger.*": Light 2005; Grewen 2005; Lim 2005.

100 "... *sexual behavior, especially in females.*": Young 2005; Cushing 2000; Gingrich 2000; Carter 1997.

100 "... *got a placebo nasal spray.*": Kosfeld 2005; Zak 2005.

100 "... *the brain's trust-circuits.*": Light 2005.

100 "... *oxytocin in the female brain.*": UvnäsMoberg 2003; Turner 1999.

100 "... *out of their brain chemicals.*": Dreher 2005.

100 "... *caution and aversion circuits.*": Carter 1998.

101 "... *loving and trusting others.*": Carter 2003, 2006 personal communication.

101 "... *harder to achieve and sustain.*": Bowlby 1980, 1988.

102 "... *by additional neurological circuits.*": Leckman 1999.

102 "... *critical judgment, lit up.*": Bartels 2000.

102 "... *pituitary and the hypothalamus.*": Insel 2004.

103 "... *primarily oxytocin and estrogen.*": Bielsky 2004; ibid Carter 2003.

103 "... *need both these neurohormones.*": Leckman 1999.

104 "... *when she isn't present.*": Lim 2004.

104 "... *woman to release oxytocin.*": Fisher 2004.

104 "... *form lifelong mating partnerships.*": Carter 1992.

105 "... *Swedish researcher Kerstin Uvnäs-Moberg.*": Uvnäs-Moberg 2001, 2004.

105 "... *pleasure, comfort, and calm.*": Taylor 2006; Depue 2005; Uvnäs-Moberg 2003.

105 "... *differences between males and females.*": Carter 1995.

105 "... *female that he finds.*": Uvnäs-Moberg 2003.

105 "... *desire when under stress.*": DeVries 1996.

106 "... *that montane voles lack.*": Young 2001.

106 "... *may account for this variability.*": Gray 2004.

106 "... *stay-at-home dads.*": Young 2005.

106 "... *fertile, and flirtatious female.*": Hammock 2005.

107 "... *match their social behaviors.*": Hammock 2005.

107 "... *have the long version of the gene.*": de Waal 2005.

107 "... *of profound social deficit.*": Wassink 2004.

107 "... *this gene and in hormones.*": Gray 2004.

107 "... *know women cheat, too.*": see Chapter 4, "Sex."

108 "... *of both men and women.*": Sabarra 2006; Aron 2005.

108 "... *about losing the beloved.*": Eisenberger 2004.

109 "... *have revealed their accuracy.*": ibid.

109 "... *dangers of social separation.*": ibid.

FOUR: SEX: THE BRAIN BELOW THE BELT

111 *". . . brain—has been deactivated."*: Holstege 2003.

112 *". . . typical man to reach orgasm."*: Carter 2006 personal communication.

112 *". . . dopamine, oxytocin, and endorphins."*: Matthews 2005; McCarthy 1996; Carter 1992.

113 *". . . threshold of orgasm easily."*: Holstege 2003.

113 *". . . felt like engaging in sex."*: ibid.

113 *". . . likely to have easy orgasms."*: Hill 2002.

113 *". . . with a new person."*: Sprecher 2002.

114 *". . . clitoris when it's aroused."*: O'Connell 2005.

115 *". . . description of the penis."*: ibid.

115 *". . . sexual enjoyment in women."*: Enserink 2005; Harris 2004.

118 *". . . allowed to fertilize her eggs."*: Eberhard 1996; Bellis 1990.

119 *". . . brains choose a winner."*: Colson 2006; Birkhead 1998.

119 *". . . had pulled off his condom."*: Singer 1973; Fox 1970.

119 *". . . not have an orgasm."*: Dawood 2005.

120 *". . . parts are more symmetrical."*: Thornhill 1999.

120 *". . . the quality of his genes."*: Fisher 2005 personal communication; Fisher 2004.

120 *". . . sexually active heterosexual couples."*: Thornhill 1995.

121 *". . . the women they date."*: ibid.

122 *". . . the use of contraception."*: Martin-Loeches 2003; Thornhill 1995.

122 *". . . female orgasm during copulation."*: Thornhill 1995.

122 *". . . right time of the month."*: Gangestad 1998.

122 *". . . imperceptible effects of male pheromones."*: Savic 2001; Grammer 1993; Getchell 1991.

122 *". . . skin and sweat glands."*: McClintock 1998, 2005.

122 *". . . leads up to ovulation."*: Getchell 1991; Gangestad 1998.

122 *". . . hour of the menstrual cycle."*: Dreher 2005; Gangestad 1998.

123 *". . . their mental focus sharpens."*: Lundstrom 2003; McClintock 2002; Savic 2001; Graham 2000.

123 *". . . certain times of the month."*: Hummel 2005; Grammer 1993.

124 *". . . pheromones and the female brain."*: Havlicek 2005.

124 *". . . and living with their mothers."*: Arnqvist 2005.

124 "... *feel certain they fathered*.": Baker 1993.

124 "... *confer a reproductive advantage*.": Hrdy, 1997.

124 "... *sperm will make it to the egg*.": Baker 1993.

124 "... *flirtatious time of the month*.": Pillsworth 2004; Buss 2002.

124 "... *with their stable partners*.": Thornhill 1995.

124 "... *semen from their secret lovers*.": Baker 1995.

124 "... *monogamy than men are*.": Hrdy 1997.

125 "... *ovaries and adrenal glands*.": Swerdloff 2002.

125 "... *to 'turn on' the brain*.": ibid.

125 "... *equivalent of sexual sassiness*.": Jenkins 2003.

125 "... *thoughts and more masturbation*.": Bancroft 1991.

125 "... *predictor of first intercourse*.": Wells 2005; Halpern 1997.

126 "... *age nine and fifteen*.": Styne 2002.

126 "... *will persist through life*.": Morris 1987.

126 "... *lose sexual interest altogether*.": see Appendix 1, "The Female Brain and Hormone Therapy."

126 "... *her with testosterone therapy*.": see Appendix 1, "The Female Brain and Hormone Therapy."

127 "... *testosterone in a woman's system*.": Pazol 2005; Krueger 2002; Schumacher 2002; Mani 2002.

127 "... *weeks of their menstrual cycle*.": Panzer 2006; Salonia 2005; Bullivant 2004; Slob 1996.

127 "... *structures in the female brain*.": see Chapter 1, "The Birth of the Female Brain."

128 "... *their most fertile days*.": Bancroft 2005; Laumann 1999, 2005; Lunde 1991.

128 "... *hypothalamus to grow larger*.": see Chapter 1, "The Birth of the Female Brain."

129 "... *sex even more than usual*.": Buss 1989.

129 "... *or perhaps another man*.": Sprecher 2002; Buss 2002.

129 "... *falling out of love with him*.": Koch 2005.

129 "... *search for the other man*.": Rilling 2004.

FIVE: THE MOMMY BRAIN

131 *". . . and in many ways, irreversibly.":* Lonstein 2005; O'Day 2001; Morgan 1992.

131 *". . . physical contact with my child.":* Soldin 2005; Stern 1989, 1993; Morgan 1992.

132 *". . . ensure their own survival.":* Martel 1993; Buntin 1984.

132 *". . . closeness with her child.":* Johns 2004; Fleming 1997; De Wied 1997.

132 *". . . daily contact with an infant.":* Lambert 2005.

132 *". . . huge increases of oxytocin.":* Fries 2005; Carter 2003; Kinsley 1999; Morgan 1992.

133 *". . . responses and priorities in life.":* Pawluski 2006; Gatewood 2005; Bodensteiner 2006; Routtenberg 2005.

133 *". . . thing from her mind.":* Story 2005.

133 *". . . job and career path.":* Xerri 1994.

134 *". . . that induces baby lust.":* McClintock 2002.

134 *". . . her fetus and placenta.":* Soldin 2005.

136 *". . . brain are flooded with them.":* Kaiser 2005; Brunton 2005; Strauss 2004.

136 *". . . feeling of stress during pregnancy.":* Kajantie 2006.

137 *". . . of her changing reality.":* Richardson 2006; Darnaudery 2004.

137 *". . . woman's brain is actually shrinking.":* Oatridge 2002.

137 *". . . months after giving birth.":* Furuta 2005.

137 *". . . flexibility of female brains.":* Kinsley 2006; Hamilton 1977.

137 *". . . metabolic changes going on.":* Holdcroft 2005.

137 *". . . networks of maternal circuits.":* Pawluski 2006.

138 *". . . uterus to start contracting.":* Mann 2005.

138 *". . . touch, sight and smell.":* Insel 2001.

139 *". . . odor of her newborn.":* Kendrick 1992.

139 *". . . about 90 percent accuracy.":* Fleming 1997; Fleming 1993.

139 *". . . tattooed on her brain.":* ibid Lonstein 2005; Pedersen 2003; Kendrick 2000.

140 *". . . and tracking her baby.":* Li 2003.

140 *". . . efficient in catching prey.":* Bodensteiner 2006; Lambert 2005.

141 "... *maintain the mommy brain.*": Bridges 2005; Featherstone 2000; Morgan 1992.

141 "... *of their pregnant mates.*": Carter 2004; Berg 2002; Storey 2000.

141 "... *pregnancy with their partners.*": Masoni 1994.

141 "... *nurturing and lactation hormone.*": Fleming 2002.

141 "... *climb higher than usual.*": Gray 2004

141 "... *cries of babies better.*": Fleming 2002.

142 "... *when a baby screams.*": Seifritz 2003.

142 "... *sex drive during this time.*": Gray 2004.

142 "... *interact with their babies.*": Storey 2000; Masoni 1994.

142 "... *mechanisms of the mommy brain.*": Sherman 2003; Neff 2003; Buchan 2003.

143 "... *in response to both photos.*": Bartels 2004.

143 "... *of exhilaration and attachement.*": Amdam 2006; Fisher 2005; Bartels 2004; Leibenluft 2004; Nitschke 2004; see Chapter 3, "Love and Trust."

143 "... *looking at a loved one.*": Bartels 2004.

143 "... *brain by estrogen and oxytocin.*": Miller 2005; Byrnes 2002.

143 "... *the first year postpartum.*": Mass 1998.

144 "... *baby is close physically.*": Uvnäs-Moberg 2003; Carter 1997; Morgan 1992.

144 "... *of cocaine every time.*": Ferris 2005.

144 "... *feelings for your baby.*": Uvnäs-Moberg 1998, 2003.

144 "... *desire for her partner.*": DeJudicibus 2002; Alder 1989; Reamy 1987.

145 "... *mellow, mildly unfocused state.*": Heinrichs 2002, 2001; Buckwalter 1999.

146 "... *at six months postpartum.*": see Appendix 2, "The Female Brain and Postpartum Depression."

146 "... *the benefits of nursing.*": Matthiesen 2001.

147 "... *relaxing waves of hormone.*": Buhimschi 2004.

147 "... *even waves of panic.*": Neighbors 2003; Uvnäs-Moberg 2003; Chezem 1997.

147 "... *brain levels of oxytocin.*": Heinrichs 2001.

147 "... *the bloodstream and brain.*": UvnäsMoberg 2003.

148 "... *circuits of their children.*": Call 1998.

148 "... *their daughters and granddaughters.*": Maestripieri 2005; Fleming 2002; Meaney 2001; Francis 1999.

148 "... *genes—type of inheritance.*": Vassena 2005; Weaver 2004; Fleming 1999.

148 "... *use estrogen and oxytocin.*": Cushing 2005; Weaver 2005; Vassena 2005; Cameron 2005; Champagne 2001, 2003; Meaney 2001.

148 "... *environment happens before puberty.*": Cameron 2005; Francis 2002.

149 "... *she mothers your grandchildren.*": Young 2005.

149 "... *in the succeeding generation.*": Gutteling 2005; Belsky 2005; Krpan 2005; Maestripieri 2005; Caldji 2000; Francis 1999.

149 "... *able to give their kids.*": Cameron 2005; Belsky 2002; Repetti 1997; Rosenblum 1994.

150 "... *of their own children.*": Francis 2002.

150 "... *and fearful as adults.*": Charmandari 2005; Lederman 2004; Darnaudery 2004; Morley-Fletcher 2004; Fleming 2002; McCormick 1999.

150 "... *stress on PET scans.*": Pruessner 2004; Hall 2004.

150 "... *from an overstressed mother.*": Francis 2002.

152 "... *other communities or species.*": Hrdy 1999.

153 "... *female siblings for care.*": Glazer 1992.

154 "... *peers in the other environments.*": Coplan 2005.

154 "... *into adolescence and adulthood.*": ibid.

154 "... *allomaternal care from others.*": Hrdy 2005; personal communication.

155 "... *context in which she mothers.*": Paris 2002.

155 "... *success as a mother.*": Taylor 1997; Fleming 1992.

SIX: EMOTION: THE FEELING BRAIN

157 "... *emotionally sensitive than men?*": Butler 2005; Wager 2005; Simon 2004; Brebner 2003; Kring 1998, 2000; Brody 1993, 1997; Briton 1995; Grossman 1993; Crawford 1992; Fagot 1989; Brody 1985; Balswick 1977; Allen 1976.

157 "... *hit him on the head?*": Samter 2002; Feingold 1994.

158 *". . . in the human male brain."*: Orzhekhovskaia 2005; Uddin 2005; Oberman 2005 and 2005 personal communication; Ohnishi 2004.

159 *". . . and emotions were hers."*: Mitchell 2005.

159 *". . . sense of this mismatch."*: Schirmer 2002, 2004, 2005.

160 *". . . innermost feelings of others."*: Brody 1985.

160 *". . . of despair and distress."*: Hall 2004.

160 *". . . that men can't ignore."*: Campbell 1993, 2005; Levenson 2003; Vingerhoets 2000; Timmers 1998; Wagner 1993; HooverDempsey 1986; Frey 1985.

160 *". . . infidelity at a gut level."*: Naliboff 2003.

160 *". . . puberty, they increase."*: Leresche 2005.

160 *". . . pain more than boys."*: Lawal 2005; Derbyshire 2002.

161 *". . . register in the body."*: Lawal 2005.

161 *". . . to brain scan studies."*: Butler 2005.

161 *". . . is grounded in biology."*: Levenson 2003.

161 *". . . integrating negative emotions."*: Butler 2005; Pujol 2002.

162 *". . . was going on in his brain."*: Rotter 1988.

162 *". . . voice for emotional nuance."*: Campbell 2005; Rosip 2004; Weinberg 1999.

162 *". . . junctures in the therapy."*: Raingruber 2001.

162 *". . . story might be feeling."*: McClure 2000; Hall 1978, 1984.

163 *". . . intuiting what they are feeling."*: Oberman 2005.

163 *". . . kind of emotional mirroring."*: Singer 2004.

163 *". . . the pain of another person."*: Singer 2004; Idiaka 2001; Zahn-Waxler 2000.

163 *". . . were being strongly shocked."*: Singer 2006; Singer 2004.

164 *". . . and protect their children."*: Taylor 2000; Campbell 1999; Bjorklund 1996; Archer 1996; Buss 1995.

164 *". . . to lose sleep than men."*: Harrison 1999.

164 *". . . electrical conductivity in the skin."*: McManis 2001; Bradley 2001; Nagy 2001; Madden 2000; Hall 2000.

164 *". . . and more rational thought."*: Naliboff 2003; Wrase 2003.

165 *". . . discomfort—to a man."*: Campbell 1993, 2005; Shoan-Golan 2004; Levenson 2003; Frey 1985.

165 *". . . interpret the emotional meaning."*: McClure 2004; Lynam 2004; Dahlen 2004; Hall 2000.

158 *". . . characteristic of Asperger's disorder.":* Campbell 2005; Lim 2005; Baron-Cohen 2002, 2004; Wang 2004; Nagy 2001; Moffitt 2001; Loeber 2001.

159 *". . . for them to tolerate.":* Campbell 1993, 2005; Levenson 2003; Frey 1985.

166 *". . . faces—than do boys.":* McClure 2000.

166 *". . . look sad or are hurt.":* BaronCohen 2004; Blair 1999; Eisenberg 1993, 1996; Faber 1994; Kochanska 1994; ZahnWaxler 1992; Eysenck 1978.

166 *". . . 90 percent of the time.":* Erwin 1992.

166 *". . . being close to someone sad.":* Mandal 1985.

166 *". . . want to do the same.":* Cross 1997.

168 *". . . remember these early days.":* Canli 2002.

168 *". . . brains of the two sexes.":* Cahill 2003.

169 *". . . men use just one side.":* Wager 2003.

169 *". . . only two lit up in men.":* Canli 2002; Shirao 2005

169 *". . . them longer than men.":* Cahill 2003, 2005; ibid Canli 2002; Bremner 2001; Seidlitz 1998; Fujita 1991.

169 *". . . not she looked sexy.":* Horgan 2004.

169 *". . . deep within the brain.":* Zald 2003; Skuse 2003.

170 *". . . activated by emotional nuance.":* Hamann 2005; Hall 2004.

170 *". . . storage about the experience.":* Phelps 2004.

170 *". . . three-dimensional, sensory snapshot.":* Phelps 2004; Giedd 1996

171 *". . . quickly as a woman can.":* Goos 2002.

171 *". . . is clearly greater in men.":* Campbell 2005; Lovell-Badge 2005; Archer 2004, 2005; Craig 2004; McGinnis 2004; Rowe 2004; Garstein 2003; Ferguson 2000; Kring 2000; Maccoby 1998; Flannery 1993.

171 *". . . relatively larger in women.":* Goldstein 2001, 2005; Gur 2002; Giedd 1996, 1997.

171 *". . . push a man's anger button.":* Campbell 2005; Sharkin 1993.

171 *". . . have short anger fuses.":* Silverman 2003.

171 *". . . response is suddenly quicker.":* Van Honk 2001.

171 *". . . don't get angry as fast.":* Giammanco 2005; Kaufman 2005; Muller 2005; Taylor 2000; Qian 2000.

172 *". . . conflict arise in a relationship.":* Parsey 2002; Ferguson 2000;

Biver 1996; Campbell 1993; Frodi 1977.

172 "... *the anterior cingulate cortex*.": Rogers 2004; Gur 2002; Goldstein 2001.

172 "... *the fear of loss or pain*.": Butler 2005.

172 "... *a trigger-tempered male*.": Campbell 2002, 2005.

173 "... *access to the coveted toy*.": Maccoby 1998.

173 "... *anticipation and verbal circuits*.": Li 2005.

173 "... *angry at a third person*.": Simon 2004.

173 "... *that a man can't match*.": Li 2005.

173 "... *frightened and shuts down*.": Calder 2001; Thunberg 2000.

174 "... *particular experiences in life*.": Butler 2005; Mcclure 2004; Wood 1998.

175 "... *anticipation of fear or pain*.": Butler 2005; Garstein 2003; Cote 2002; Nagy 2001; Brody 1985; Carey 1978.

175 "... *pleasure/reward circuits fire*.": Etkin 2006 personal communication; Rogan 2005.

175 "... *of danger or pain*.": Butler 2005.

175 "... *more common in women*.": Antonijevic 2006; Halbreich 2006; Simon 2004; Johnston 1991.

175 "... *through their reproductive years*.": Madden 2000.

175 "... *women's risk of depression*.": Kendler 2006.

175 "... *turned on by estrogen*.": Lee 2005; Abraham 2005.

175 "... *run in certain families*.": Altshuler 2005.

175 "... *threats and severe stress*.": Staley 2006; Pezawas 2005; Bertolino 2005; Halari 2005; Kaufman 2004; Barr 2004; Caspi 2003; Auger 2001.

175 "... *into a clinical depression*.": Bertolino 2005.

175 "... *chemical or hormonal rebalancing*.": Staley 2006; Altshuler 2001; Jensvold 1996.

SEVEN: THE MATURE FEMALE BRAIN

181 "... *certain times of the month*.": Protopopescu 2005; Morgan 2004.

181 "... *flood her so quickly anymore*.": Labouvie-Vief 2003.

181 "... *oxytocin is now down too*.": Yamamoto 2006; Taylor 2006; Light 2005; Matthews 2005; Morgan 2004.

181 "... *attentive to others' personal needs.*": Light 2005; Tang 2005.

181 "... *help those around her.*": Light 2005.

182 "... *who we ought to be.*": Winfrey 2005.

183 "... *tending hormone—also dropped.*": Yamamoto 2006; Light 2005; Motzer 2004; Tang 2003.

183 "... *her newly found balance.*": Protopopsecu 2005; Motzer 2004; Morgan 2004; Labouvie-Vief 2003.

183 "... *functioning of their brains.*": Kirsch 2005; Tang 2005; Windle 2004.

183 "... *pain to anxiety and depression.*": Soares 2000, 2001, 2003, 2004, 2005; Schmidt 2004.

183 "... *sensitivity in the brain itself.*": Weiss 2004.

183 "... *rocket fuel for sex drive.*": Burger 2002.

184 "... *a rocky ride for some women.*": see Appendix 1, "The Female Brain and Hormone Therapy."

184 "... *experience estrogen withdrawal.*": Lobo 2000.

185 "... *cravings for sweets and carbs.*": Ratka 2005; Joffe 1998, 2002, 2003.

185 "... *PMS and postpartum hypothyroidism.*": Duval 1999.

186 "... *already passed through menopause.*": Guthrie 2003; see Appendix 1, "The Female Brain and Hormone Therapy."

186 "... *plummet at this time too.*": Burger 2002; see Appendix 1, "The Female Brain and Hormone Therapy."

186 "... *more extreme than most.*": Davison 2005.

186 "... *it was barely present.*": Davis 2005.

186 "... *slapped it on that very day.*": Braunstein 2005; Bolour 2005; Goldstat 2003; Shifren 2000.

187 "... *less frequent and intense.*": Laumann 1999, 2005; see Chapter 4, "Sex" and Appendix 1, "The Female Brain and Hormone Therapy."

187 "... *had at age twenty.*": Davison 2005.

187 "... *gels—that are now available.*": Wang 2004; Shifren 2000.

187 "... *or their sexual interest.*": Davis 2005.

189 "... *are now set free.*": Taylor 2006.

190 "... *pups died as a result.*": Stern 1989, 1993; Morgan 1992.

191 "... *is important to them.*": Shellenbarger 2005.

192 "... *she was really mad.*": Helson 1992.

193 *". . . more like a man's."*: Lobo 2000.

193 *". . . interaction of their relationship."*: Swaab 1995, 2001; Kruijver 2001; Fernandez-Guasti 2000.

193 *". . . during the menopause years."*: Kiecolt-Glasser 2005; Mackey 2001; Robinson 2001.

194 *". . . rules of their relationship."*: Sbarra 2006; Kruijver 2001.

194 *". . . fifty are initiated by women."*: U.S. Human Resources Services Administration 2002.

195 *". . . in arguing with Robert."*: Seeman 2001; Gust 2000; Burleson 1998.

196 *". . . nurturing instincts had returned."*: Taylor 2006; Miller 2002.

196 *". . . stress response is kept low."*: Kajantie 2006; Morgan 2004.

198 *". . . or had her children."*: Helson 2005.

198 *". . . higher than did other women."*: Helson 2001, 2005; Roberts 2002.

199 *". . . own work and activities."*: Kiecolt-Glaser 1996, 1998.

200 *". . . on estrogen soon shrivel."*: Taylor 2006; McEwen 2001, 2005.

200 *". . . soon after the operation."*: Sherwin 2005.

200 *". . . stroke than men their age."*: Stirone 2005.

200 *". . . typical of much older women."*: Shaywitz 2003.

201 *". . . skills and emotional processing."*: Erickson 2005.

201 *". . . effects on the brain."*: Rossouw 2002.

201 *". . . vessels is likely gone."*: Saenz 2005; Tessitore 2005; Clarkson 2005; Brownley 2004.

201 *". . . protect brain function as well."*: Sherwin 2005.

201 *". . . specialists in hormone therapy."*: Hickey 2005; Davis 2005

202 *". . . cortex sooner than women."*: Kochunov 2005; Sullivan 2004. Li 2005.

202 *". . . no longer bear children."*: Finch 2002.

203 *". . . many ancient human populations."*: Hawkes 1998, 2004.

203 *". . . their chances of survival."*: Hawkes 2003.

203 *". . . in India and Africa."*: Beise 2002.

203 *". . . presence of a father."*: Hawkes 2003.

204 *". . . lot of years to imagine."*: Kenyon 2005; Arantes-Oliveira 2003; Murphy 2003; Wise 2003.

Notes

APPENDIX ONE: THE FEMALE BRAIN AND HORMONE THERAPY

213 "... *middle have felt betrayed*.": Ekstrom 2005; Hickey 2005.

214 "... *and off hormone therapy?*": Brownley 2004.

214 "... *laboratory rodents and primates*.": Wise 2005; Clarkson 2005; Papalexi 2005.

214 "... *function as we age*.": Hultcrantz 2006; Erickson 2005; Saenz 2005; Murabito 2005; Zemlyak 2005.

214 "... *for some emotional processing*.": Erickson 2005; Shaywitz 2003

214 "... *of WHI and WHIMS*.": Franklin 2006; ibid Erickson 2005; Li 2005; Gulinello 2005; Stirone 2005.

215 "... *due sometime after 2010*.": Harman 2004, 2005.

215 "... *related to verbal memory*.": Resnick 2001; Maki 2001.

215 "... *shrinkage seen with age*.": Raz 2004.

215 "... *age than women's brains*.": Kochunov 2005.

215 "... *area involved in facial recognition*.": Raz 2004; Sullivan 2004.

215 "... *they outperformed men, too*.": Miller 2002.

215 "... *women who took HT*.": Erickson 2005; Raz 2004; ibid.

215 "... *but from other women's*.": Murabito 2005.

216 "... *timing of the hormone treatment*.": Rasgon 2005.

216 "... *estrogen therapy following menopause*.": Sherwin 2005; Rubinow 2005; Wise 2005; Turgeon 2004.

217 "... *before the day of menopause*.": Burger 2002; Lobo 2000.

217 "... *less sensitive to estrogen*.": Weiss 2004.

217 "... *two years before menstruation stops*.": Soares 2004, 2005; Schmidt 2005; Rasgon 2005; Douma 2005.

217 "... *cells—have become disturbed*.": Bethea 2005.

217 "... *alone if it is mild*.": Bertschy 2005; Rubinow 2002; Schmidt 2000; Komesaroff 1999.

218 "... *estrogen and stress sensitivity*.": Kajantie 2006; Morgan 2004; Seeman 2001; Gust 2000; Burleson 1998.

218 "... *serotonin, norepinephrine, and dopamine*.": Tessitore 2005.

218 "... *with or without hot flashes*.": Kravitz 2005; Joffe 2002.

218 "... *you're over age forty*.": Kravitz 2005.

218 "... *or more past menopause*.": Guthrie 2005; Joffe 2002; Henderson

2002; Dennerstein 1997, 2000.

219 *". . . first five years after menopause."*: Davis 2005; Erickson 2005; McEwen 2005; Sherwin 2005; Shaywitz 2003; Woolley 2002; Cummings 2002; Halbreich 1995; Craik 1977.

219 *". . . room or even before surgery."*: Sherwin 2005.

219 *". . . makes them feel smart again."*: Korol 2004; Farr 2000.

220 *". . . during the menopausal transition."*: Wright 2004.

220 *". . . scientific rigor that they deserve."*: Naftolin 2005.

220 *". . . should not start HT."*: Clarkson 2005.

220 *". . . estrogenonly therapy whenever possible."*: Lobo 2005; Speroff 2005.

221 *". . . heart, uterus, and vascular system."*: Mendelsohn 2005.

221 *". . . stress reduction and meditation practice."*: Perez-Martin 2005; Bough 2005; Mogi 2005; Yonezawa 2005; Gulati 2005; Elavsky 2005; Hickey 2005; Davison 2005; Brizendine 2004; Epel 2004.

221 *". . . gels, injectables, or pellets."*: Goldstat 2003.

221 *". . . exercise and relaxation therapies."*: Brizendine 2004.

222 *". . . are the cause of weight gain."*: Bakken 2004.

224 *". . . not help slow it down."*: Morse 2005.

224 *". . . than men their age."*: Stirone 2005.

224 *". . . greater brain cell death."*: Sastre 2002.

224 *". . . women live longer than men."*: Vina 2005.

224 *". . . cables between brain areas."*: Henderson 2002.

224 *". . . slowly or not at all."*: Tanapat 2002.

225 *". . . for developing Alzheimer's disease."*: Alvarez 2005.

225 *". . . deficient levels of estrogen."*: Yue 2005; Li 2005.

225 *". . . after menopause offers no benefit."*: Woods 2000.

225 *". . . more benefit from social support."*: Kajantie 2006; Epel 2006; Gurung 2003.

225 *". . . exercises, such as playing cards."*: Podewils 2005.

226 *". . . percent of their testosterone."*: Davis 2005; Braunstein 2005; Burger 2002; Shifren 2000.

226 *". . . hormone transition called 'adrenopause'."*: Nawata 2004.

226 *". . . testes when they were young."*: Vermeulen 1995.

226 *". . . to 70 for women."*: Lobo 2000.

227 *". . . to maintain sexual interest."*: Gray 1991.

227 "... *very low testosterone levels.*": Guay 2004.

227 "... *climbs to six in ten.*": Laumann 1999.

227 "... *patterns all over the world.*": Laumann 2005.

227 "... *when we were twenty.*": Gray 1991.

228 "... *had their ovaries removed.*": Laumann 1999, 2005.

228 "... *creams or pills may help.*": Warnock 2005.

228 "... *women should feel entitled.*": see Chapter 4, "Sex."

229 "... *the course of a woman's life.*": Basson 2005.

299 "... *psychological 'resistance' to sex.*": Basson 2005.

230 "... *sexual interest soon returned.*": Sherwin 1985.

230 "... *creams for women patients.*": Guay 2002; Bachmann 2002.

230 "... *sexual interest back to par.*": Sherwin 1985.

230 "... *her desire for partner sex.*": Apperloo 2003; Davis 1998, 2001.

230 "... *sexual interest in all women.*": Buster 2005; Davison 2005.

231 "... *a cause of sexual dysfunction.*": Guay 2004.

231 "... *both sexes with testosterone therapy.*": Davison 2005; Connell 2005; Guay 2002.

231 "... *to assume the downside risks.*": Rhoden 2004; Wang 2004; Rossouw 2002.

APPENDIX TWO: THE FEMALE BRAIN AND POSTPARTUM DEPRESSION

233 "... *most common is postpartum depression.*": Logsdon 2006; Zonana 2005; Brandes 2004.

233 "... *the relevant hormonal changes.*": Hasser 2006; Kendler 2006; Boyd 2006.

234 "... *women with postpartum depression.*": Bloch 2003, 2006.

234 "... *the stress hormone cortisol.*": Bloch 2005.

234 "... *high stress in the home.*": O'Hara 1991.

235 "... *sleep, the child, and the partner.*": Edhborg 2005.

235 "... *postpartum depression in certain women.*": Uvnäs-Moberg 2003.

235 "... *development of postpartum depression.*": Walker 2004.

235 "... *such as supportive talk therapy.*": Magalhaes 2006; Altshuler 2001.

APPENDIX THREE: THE FEMALE BRAIN AND SEXUAL ORIENTATION

237 *". . . percent of the female population."*: Jorm 2003.

237 *". . . different than in men."*: Rahman 2005.

237 *". . . reporting more bisexual interests."*: Bocklandt 2006; Rahman 2005; Chivers 2004; Sandfort 2003.

237 *". . . woman than a gay man."*: Sandfort 2003.

237 *". . . matter of brain wiring."*: LeVay 1991.

237 *". . . and female sexual orientation."*: Mustanski 2005; Pattatucci 1995; Pillard 1995.

238 *". . . tumble play and sexual attraction."*: Hershberger 2004.

238 *". . . not exposed to fetal testosterone."*: Hines 2004; Manning 2004; see Chapter 1, "The Birth of the Female Brain."

238 *". . . between straight and gay women."*: Rahman 2003.

238 *". . . auditory response—a male-typical pattern."*: McFadden 1998, 1999.

238 *". . . intermediate between males and females."*: Muscarella 2004.

238 *". . . their gay female counterparts."*: Rahman 2003.

REFERENCES

Abraham, I. M., and A. E. Herbison (2005). "Major sex differences in nongenomic estrogen actions on intracellular signaling in mouse brain in vivo." *Neuroscience* 131 (4): 945–51.

Adams, D. (1992). "Biology does not make men more aggressive than women." In K. Bjorkqvist and P. Niemela, eds., *Of mice and women: Aspects of female aggression,* 17–26. San Diego: Academic Press.

Adler, E. M., A. Cook, et al. (1986) "Hormones, mood and sexuality in lactating women." *Br J Psychiatry* 148:74–79.

Agrati, D., A. Fernandez-Guasti, et al. (2005). "Compulsive-like behaviour according to the sex and the reproductive stage of female rats." *Behav Brain Res* 161 (2): 313–19.

Alder, E. M. (1989). "Sexual behaviour in pregnancy, after childbirth and during breastfeeding." *Baillieres Clin Obstet Gynaecol 3* (4): 805–21.

Alele, P. E., and L. L. Devaud (2005). "Differential adaptations in GABAergic and glutamatergic systems during ethanol withdrawal in male and female rats." *Alcohol Clin Exp Res* 29 (6): 1027–34.

Alexander, G. M., B. B. Sherwin, et al. (1990). "Testosterone and sexual behavior in oral contraceptive users and nonusers: A prospective study." *Horm Behav* 24 (3): 388–402.

Allen, J. (1976). "Sex differences in emotionality." *Human Relations* 29:711–22.

Altemus, M., L. Redwine, et al. (1997). "Reduced sensitivity to glucocorticoid feedback and reduced glucocorticoid receptor mRNA expression in the luteal phase of the menstrual cycle." *Neuropsychopharmacology* 17 (2): 100–109.

Altemus, M., C. Roca, et al. (2001). "Increased vasopressin and adrenocorticotropin responses to stress in the midluteal phase of the menstrual cycle." *J Clin Endocrinol Metab* 86 (6): 2525–30.

Altemus, M., and E. Young (2006). "The menstrual cycle and cortisol feedback sensitivity with metyrapone." In preparation.

Altshuler, D., L. D. Brooks, et al. (2005). "A haplotype map of the human genome." *Nature* 437 (7063): 1299–320.

Altshuler, L. L., L. S. Cohen, et al. (2001). "The expert consensus guideline series: Treatment of depression in women." *Postgrad Med* (Spec. No.): 1–107.

Altshuler, L. L., L. S. Cohen, et al. (2001). "Treatment of depression in women: A summary of the expert consensus guidelines." *J Psychiatr Pract* 7 (3): 185–208.

Alvarez, D. E., I. Silva, et al. (2005). "Estradiol prevents neural tau hyperphosphorylation characteristic of Alzheimer's disease." *Ann NY Acad Sci* 1052:210–24.

Amdam, G. V., A. Csondes, et al. (2006). "Complex social behaviour derived from maternal reproductive traits." *Nature* 439 (7072): 76–78.

Antonijevic, I. (2006). "Depressive disorders—is it time to endorse different pathophysiologies?" *Psychoneuroendocrinology* 31 (1): 1–15.

Apperloo, M. J., J. G. Van Der Stege, et al. (2003). "In the mood for sex: The value of androgens." *J Sex Marital Ther* 29 (2): 87–102; discussion 177–79.

Arantes-Oliveira, N., J. R. Berman, et al. (2003). "Healthy animals with extreme longevity." *Science* 302 (5645): 611.

Archer, J. (1991). "The influence of testosterone on human aggression." *Br J Psychol* 82 (Pt. 1): 1–28.

Archer, J. (1996). "Sex differences in social behavior: Are the social role and evolutionary explanations compatible?" *American Psychologist* 51 (9): 909–17.

REFERENCES

Archer, J. (2004). "Sex differences in aggression in real-world settings: A meta-analytic review." *Review of General Psychology* 8:291–322.

Archer, J. C., (2005). "An integrated review of indirect, relational, and social aggression." *Personality and Social Psychology Review* 9 (3): 212–30.

Arnold, A. P. (2004). "Sex chromosomes and brain gender." *Nat Rev Neurosci* 5 (9): 701–8.

Arnold, A. P., and P. S. Burgoyne (2004). "Are XX and XY brain cells intrinsically different?" *Trends Endocrinol Metab* 15 (1): 6–11.

Arnold, A. P., J. Xu, et al. (2004). "Minireview: Sex chromosomes and brain sexual differentiation." *Endocrinology* 145 (3): 1057–62.

Arnqvist, G., and M. Kirkpatrick (2005). "The evolution of infidelity in socially monogamous passerines: The strength of direct and indirect selection on extrapair copulation behavior in females." *Am Nat* 165 (Suppl. 5): S26–37.

Arnsten, A. F., and R. M. Shansky (2004). "Adolescence: Vulnerable period for stress-induced prefrontal cortical function? Introduction to part IV." *Ann NY Acad Sci* 1021:143–47.

Aron, A., H. Fisher, et al. (2005). "Reward, motivation, and emotion systems associated with early-stage intense romantic love." *J Neurophysiol* 94 (1): 327–37.

Auger, A. P., D. P. Hexter, et al. (2001). "Sex difference in the phosphorylation of cAMP response element binding protein (CREB) in neonatal rat brain." *Brain Res* 890 (1): 110–17.

Azurmendi, A., F. Braza, et al. (2005). "Cognitive abilities, androgen levels, and body mass index in 5-year-old children." *Horm Behav* 48 (2): 187–95.

Babcock, S., and S. Laschever (2004). *Women don't ask: Negotiation and the gender Divide.* Princeton: Princeton University Press.

Bachevalier, J., M. Brickson, et al. (1990). "Age and sex differences in the effects of selective temporal lobe lesion on the formation of visual discrimination habits in rhesus monkeys (*Macaca mulatta*)." *Behav Neurosci* 104 (6): 885–99.

Bachevalier, J., C. Hagger, et al. (1989). "Gender differences in visual habit formation in 3-month-old rhesus monkeys." *Dev Psychobiol* 22 (6): 585–99.

Bachevalier, J., and C. Hagger (1991). "Sex differences in the development of learning abilities in primates." *Psychoneuroendocrinology* 16 (1–3): 177–88.

Bachmann, G., J. Bancroft, et al. (2002). "Female androgen insufficiency: The Princeton consensus statement on definition, classification, and assessment." *Fertil Steril* 77 (4): 660–65.

Baker, R., and M. A. Bellis (1993). "Human sperm competition: Ejaculate adjustment by males and the function of masturbation, non-paternity rates." *Animal Behaviour* 46 (5): 861–65.

Baker, R., and M. A. Bellis (1993). "Human sperm competition: Ejaculate manipulation by females and a function for the female orgasm." *Animal Behaviour* 46 (5): 887–909.

Baker, R., and M. A. Bellis (1995) *Human Sperm Competition: Copulation, Masturbation, and Infidelity*. London and New York: Chapman & Hall.

Bakken, K., A. E. Eggen, et al. (2004). "Side-effects of hormone replacement therapy and influence on pattern of use among women aged 45–64 years: The Norwegian Women and Cancer (NOWAC) study 1997." *Acta Obstet Gynecol Scand* 83 (9): 850–56.

Balswick, J. (1977). "Differences in expressiveness: Gender." *Journal of Marriage and the Family* 39:121–27.

Bancroft, J. (2005). "The endocrinology of sexual arousal." *J Endocrinol* 186 (3): 411–27.

Bancroft, J., and D. Rennie (1993). "The impact of oral contraceptives on the experience of perimenstrual mood, clumsiness, food craving and other symptoms." *J Psychosom Res* 37 (2): 195–202.

Bancroft, J., B. B. Sherwin, et al. (1991). "Oral contraceptives, androgens, and the sexuality of young women: I. A comparison of sexual experience, sexual attitudes, and gender role in oral contraceptive users and nonusers." *Arch Sex Behav* 20 (2): 105–20.

Baron-Cohen, S. (2002). "The extreme male brain theory of autism." *Trends Cogn Sci* 6 (6): 248–54.

Baron-Cohen, S., and M. K. Belmonte (2005). "Austism: A window onto the development of the social and the analytic brain." *Annu Rev Neurosci* 28:109–26.

Baron-Cohen, S., and Bruce J. Ellis (ED) (2005). "The empathizing system: A revision of the 1994 model of the mindreading system." In *Origins of the Social Mind: Evolutionary Psychology and Child*

References

Development, 468–92. New York: Guilford Press.

Baron-Cohen, S., R. C. Knickmeyer, et al. (2005). "Sex differences in the brain: Implications for explaining autism." *Science* 310 (5749): 819–23.

Baron-Cohen, S., J. Richler, et al. (2003). "The systemizing quotient: An investigation of adults with Asperger syndrome or high-functioning autism, and normal sex differences." *Philos Trans R Soc Lond B Biol Sci* 358 (1430): 361–74.

Baron-Cohen, Simon, et al. (2004). *Prenatal testosterone in mind: Amniotic fluid studies.* Cambridge, MA: MIT Press.

Baron-Cohen, S., and S. Wheelwright (2004). "The empathy quotient: An investigation of adults with Asperger syndrome or high functioning autism, and normal sex differences." *J Autism Dev Disord* 34 (2): 163–75.

Barr, C. S., T. K. Newman, et al. (2004). "Early experience and sex interact to influence limbic-hypothalamic-pituitary-adrenal-axis function after acute alcohol administration in rhesus macaques (*Macaca mulatta*)." *Alcohol Clin Exp Res* 28 (7): 1114–19.

Barr, C. S., T. K. Newman, et al. (2004). "Interaction between serotonin transporter gene variation and rearing condition in alcohol preference and consumption in female primates." *Arch Gen Psychiatry* 61 (11): 1146–52.

Barr, C. S., T. K. Newman, et al. (2004). "Sexual dichotomy of an interaction between early adversity and the serotonin transporter gene promoter variant in rhesus macaques." *Proc Natl Acad Sci USA* 101 (33): 12358–63.

Bartels, A., and S. Zeki (2000). "The neural basis of romantic love." *Neuroreport* 11 (17): 3829–34.

Bartels, A., and S. Zeki (2004). "The neural correlates of maternal and romantic love." *Neuroimage* 21 (3): 1155–66.

Bartzokis, G., and L. Altshuler (2005). "Reduced intracortical myelination in schizophrenia." *Am J Psychiatry* 162 (6): 1229–30.

Basson, R. (2005). "Women's sexual dysfunction: Revised and expanded definitions." *CMAJ* 172 (10): 1327–33.

Baumeister, R. F. (2000). "Differences in erotic plasticity: The female sex drive as socially flexible and responsive." *Psychol Bull* 126: 347–74.

Baumeister, R. F., and K. L. Sommer (1997). "What do men want? Gender differences and two spheres of belongingness: Comment on Cross and Madson (1997)." *Psychol Bull* 122 (1): 38–44; discussion 51–55.

Bayliss, A. P., G. di Pellegrino, et al. (2005). "Sex differences in eye gaze and symbolic cueing of attention." *Q J Exp Psychol A* 58 (4): 631–50.

Bayliss, A. P., and S. P. Tipper (2005). "Gaze and arrow cueing of attention reveals individual differences along the autism spectrum as a function of target context." *Br J Psychol* 96 (Pt. 1): 95–114.

Bebbington, P. (1996). "The origin of sex difference in depressive disorder: Bridging the gap." *Int Review of Psychiatry* 8:295–332.

Becker, J. B., A. P. Arnold, et al. (2005). "Strategies and methods for research on sex differences in brain and behavior." *Endocrinology* 146 (4): 1650–73.

Beem, A. L., E. J. Geus, et al. (2006). "Combined linkage and association analyses of the 124-bp allele of marker D2S2944 with anxiety, depression, neuroticism and major depression." *Behav Genet* In press.

Behan, M., and C. F. Thomas (2005). "Sex hormone receptors are expressed in identified respiratory motoneurons in male and female rats." *Neuroscience* 130 (3): 725–34.

Beise, J., and E. Voland (2002). "Effect of producing sons on maternal longevity in premodern populations." *Science* 298 (5592): 317; author reply 317.

Bell, E. C., M. C. Willson, et al. (2006). "Males and females differ in brain activation during cognitive tasks." *Neuroimage.* In press.

Bellis, M. A., R. R. Baker, et al. (1990). "A guide to upwardly mobile spermatozoa." *Andrologia* 22 (5): 397–99.

Belsky, J. (2002). "Developmental origins of attachment styles." *Attach Hum Dev* 4 (2): 166–70.

Belsky, J. (2002). "Quantity counts: Amount of child care and children's socio-emotional development." *J Dev Behav Pediatr* 23 (3): 167–70.

Belsky, J., and R. M. Fearon (2002). "Early attachment security, subsequent maternal sensitivity, and later child development: Does continuity in development depend upon continuity of caregiving?"

Attach Hum Dev 4 (3): 361–87.

Belsky, J., S. R. Jaffee, et al. (2005). "Intergenerational transmission of warm-sensitive-stimulating parenting: A prospective study of mothers and fathers of 3-year-olds." *Child Dev* 76 (2): 384–96.

Bennett, D. S., P. J. Ambrosini, et al. (2005). "Gender differences in adolescent depression: Do symptoms differ for boys and girls?" *J Affect Disord* 89 (1–3): 35–44.

Berenbaum, S. A. (1999). "Effects of early androgens on sex-typed activities and interests in adolescents with congenital adrenal hyperplasia." *Horm Behav* 35 (1): 102–10.

Berenbaum, S. A. (2001). "Cognitive function in congenital adrenal hyperplasia." *Endocrinol Metab Clin North Am* 30 (1): 173–92.

Berenbaum, S. A., and J. M. Bailey (2003). "Effects on gender identity of prenatal androgens and genital appearance: Evidence from girls with congenital adrenal hyperplasia." *J Clin Endocrinol Metab* 88 (3): 1102–6.

Berenbaum, S. A., K. Korman Bryk, et al. (2004). "Psychological adjustment in children and adults with congenital adrenal hyperplasia." *J Pediatr* 144 (6): 741–46.

Berenbaum, S. A., and D. E. Sandberg (2004). "Sex determination, differentiation, and identity." *N Engl J Med* 350 (21): 2204–6; author reply 2204–6.

Berg, S. J., and K. E. Wynne-Edwards (2002). "Salivary hormone concentrations in mothers and fathers becoming parents are not correlated." *Horm Behav* 42 (4): 424–36.

Berkley, K. (2002). "Pain: Sex/Gender differences." In *Hormones, Brain and Behavior*, ed. D. W. Pfaff, vol. 5, 409–42. San Diego: Academic Press.

Bertolino, A., G. Arciero, et al. (2005). "Variation of human amygdala response during threatening stimuli as a function of 5'HTTLPR genotype and personality style." *Biol Psychiatry* 57 (12): 1517–25.

Bertschy, G., D. De Ziegler, et al. (2005). "[Mood disorders in perimenopausal women: Hormone replacement or antidepressant therapy?]." *Rev Med Suisse* 1 (33): 2155–56, 2159–61.

Bethea, C. L., F. K. Pau, et al. (2005). "Sensitivity to stress-induced reproductive dysfunction linked to activity of the serotonin system." *Fertil Steril* 83 (1): 148–55.

Bethea, C. L., J. M. Streicher, et al. (2005). "Serotonin-related gene expression in female monkeys with individual sensitivity to stress." *Neuroscience* 132 (1): 151–66.

Bielsky, I. F., S. B. Hu, et al. (2004). "Profound impairment in social recognition and reduction in anxiety-like behavior in vasopressin V1a receptor knockout mice." *Neuropsychopharmacology* 29 (3): 483–93.

Bielsky, I. F., and L. J. Young (2004). "Oxytocin, vasopressin, and social recognition in mammals." *Peptides* 25 (9): 1565–74.

Birkhead, T. W., and A. P. Moller, eds. (1998). *Sperm Competition and Sexual Selection.* San Diego: Academic Press.

Birzniece V., T. Backstrom et. al. (2006). "Neuroactive steroid effects on cognitive functions with a focus on the serotonin and GABA systems." *Brain Res Rev.* In press.

Biver, F., F. Lotstra, et al. (1996). "Sex difference in 5HT2 receptor in the living human brain." *Neurosci Lett* 204 (1–2): 25–28.

Bjorklund, D. F., and K. Kipp (1996). "Parental investment theory and gender differences in the evolution of inhibition mechanisms." *Psychol Bull* 120 (2): 163–88.

Blair, R. J., J. S. Morris, et al. (1999). "Dissociable neural responses to facial expressions of sadness and anger." *Brain* 122 (Pt. 5): 883–93.

Blehar, M. C. (2003). "Public health context of women's mental health research." *Psychiatr Clin North Am* 26 (3): 781–99.

Blehar, M. C., and G. P. Keita (2003). "Women and depression: A millennial perspective." *J Affect Disord* 74 (1): 1–4.

Blinkhorn, S. (2005). "Intelligence: A gender bender." *Nature* 438 (7064): 31–32.

Bloch, M., R. C. Daly, et al. (2003). "Endocrine factors in the etiology of postpartum depression." *Compr Psychiatry* 44 (3): 234–46.

Bloch, M., N. Rotenberg, et al. (2006). "Risk factors for early postpartum depressive symptoms." *Gen Hosp Psychiatry* 28 (1): 3–8.

Bloch, M., D. R. Rubinow, et al. (2005). "Cortisol response to ovine corticotropin-releasing hormone in a model of pregnancy and parturition in euthymic women with and without a history of postpartum depression." *J Clin Endocrinol Metab* 90 (2): 695–99.

Bloch, M., P. J. Schmidt, et al. (2000). "Effects of gonadal steroids in women with a history of postpartum depression." *Am J Psychiatry* 157 (6): 924–30.

Bocklandt, S., S. Horvath, et al. (2006). "Extreme skewing of X chromosome inactivation in mothers of homosexual men." *Hum Genet* 118 (6): 691–94.

Bodensteiner, K. J., P. Cain, et al. (2006). "Effects of pregnancy on spatial cognition in female Hooded Long-Evans rats." *Horm Behav* 49 (3): 303–14.

Boehm, U., Zhihua Zou, and Linda Buck (2005). "GNRH cell circuitry: The brain is broadly wired for reproduction." *Cell* 123 (4): 683–95.

Bolour, S., and G. Braunstein (2005). "Testosterone therapy in women: A review." *Int J Impot Res* 17 (5): 399–408.

Bond, A. J., J. Wingrove, et al. (2001). "Tryptophan depletion increases aggression in women during the premenstrual phase." *Psychopharmacology* (Berl) 156 (4): 477–80.

Booth, A., D. R. Johnson, et al. (2003). "Testosterone and child and adolescent adjustment: The moderating role of parent-child relationships." *Dev Psychol* 39 (1): 85–98.

Born, L., A. Shea, et al. (2002). "The roots of depression in adolescent girls: Is menarche the key?" *Curr Psychiatry Rep* 4 (6): 449–60.

Bosch, O. J., S. A. Kromer, et al. (2006). "Prenatal stress: Opposite effects on anxiety and hypothalamic expression of vasopressin and corticotropin-releasing hormone in rats selectively bred for high and low anxiety." *Eur J Neurosci* 23 (2): 541–51.

Botwin, M. D., D. M. Buss, et al. (1997). "Personality and mate preferences: Five factors in mate selection and marital satisfaction." *J Pers* 65 (1): 107–36.

Bough, K., (2005). "High-fat, calorie restricted ketogenic diet, KD, stabilizes brain and increases neuron stability." Society for Neuroscience meeting, Washington, D.C.

Bowlby, J. (1980). *Attachment and Loss*, vol. 3. London: Hogarth Press.

Bowlby, J. (1988). *A Secure Base: Parent-Child Attachment and Healthy Human Development*. New York: Basic Books.

Bowman, R. E., D. Ferguson, et al. (2002). "Effects of chronic restraint stress and estradiol on open field activity, spatial memory, and monoaminergic neurotransmitters in ovariectomized rats." *Neuroscience* 113 (2): 401–10.

Boyd, R. C., L. H. Zayas, et al. (2006). "Mother-infant interaction, life events and prenatal and postpartum depressive symptoms among urban minority women in primary care." *Matern Child Health J*: In press.

Bradley, M. M., M. Codispoti, et al. (2001). "Emotion and motivation II: Sex differences in picture processing." *Emotion* 1 (3): 300–19.

Bradley, M. M., B. Moulder, et al. (2005). "When good things go bad: The reflex physiology of defense." *Psychol Sci* 16 (6): 468–73.

Brandes, M., C. N. Soares, et al. (2004). "Postpartum onset obsessive-compulsive disorder: Diagnosis and management." *Arch Women Ment Health* 7 (2): 99–110.

Braunstein, G. D., D. A. Sundwall, et al. (2005). "Safety and efficacy of a testosterone patch for the treatment of hypoactive sexual desire disorder in surgically menopausal women: A randomized, placebo-controlled trial." *Arch Intern Med* 165 (14): 1582–89.

Brebner, J. (2003). "Gender and emotions." *Personality and Individual Differences* 34:387–94.

Bremner, J. D., R. Soufer, et al. (2001). "Gender differences in cognitive and neural correlates of remembrance of emotional words." *Psychopharmacol Bull* 35 (3): 55–78.

Bridges, R. S., and V. F. Scanlan (2005). "Maternal memory in adult, nulliparous rats: Effects of testing interval on the retention of maternal behavior." *Dev Psychobiol* 46 (1): 13–18.

Briton, N. J., and J. A. Hall (1995). "Beliefs about female and male nonverbal communication." *Sex Roles* 32:79–90.

Brizendine, L. (2004). "Menopause-related depression and low libido: Fine-tuning treatment." *OBGYN Management*, 16 (8): 29–42.

Brody, L. (1997). "Gender and emotions: Beyond stereotypes." *Journal of Social Issues* 53:369–94.

Brody, L., and J. A. Hall (1993). "Gender and emotion." In M. Lewis and J. Haviland, eds., *Handbook of Emotions*, 447–60. New York: Guilford Press.

Brody, L. R. (1985). "Gender differences in emotional development: A review of theories and research." *J Pers* 53:102–49.

Brown, L. (2005). Personal communication.

Brown, W. M., L. Cronk, et al. (2005). "Dance reveals symmetry especially in young men." *Nature* 438 (7071): 1148–50.

References

Brownley, K. A., A. L. Hinderliter, et al. (2004). "Cardiovascular effects of 6 months of hormone replacement therapy versus placebo: Differences associated with years since menopause." *Am J Obstet Gynecol* 190 (4): 1052–58.

Brunton, P. J., S. L. Meddle, et al. (2005). "Endogenous opioids and attenuated hypothalamic-pituitary-adrenal axis responses to immune challenge in pregnant rats." *J Neurosci* 25 (21): 5117–26.

Buchan, J. C., S. C. Alberts, et al. (2003). "True paternal care in a multi-male primate society." *Nature* 425 (6954): 179–81.

Buckwalter, J. G., F. Z. Stanczyk, et al. (1999). "Pregnancy, the post-partum, and steroid hormones: Effects on cognition and mood." *Psychoneuroendocrinology* 24 (1): 69–84.

Buhimschi, C. S. (2004). "Endocrinology of lactation." *Obstet Gynecol Clin North Am* 31 (4): 963–79.

Bullivant, S. B., S. A. Sellergren, et al. (2004). "Women's sexual experience during the menstrual cycle: Identification of the sexual phase by noninvasive measurement of luteinizing hormone." *J Sex Res* 41 (1): 82–93.

Buntin, J. D., S. Jaffe, et al. (1984). "Changes in responsiveness to newborn pups in pregnant, nulliparous golden hamsters." *Physiol Behav* 32 (3): 437–39.

Burbank, V. K. (1987). "Female aggression in cross-cultural perspective." *Behavior Science Research* 21:70–100.

Burger, H. G., E. Dudley, et al. (2002). "The ageing female reproductive axis I." *Novartis Found Symp* 242:161–67; discussion 167–71.

Burger, H. G., E. C. Dudley, et al. (2002). "Hormonal changes in the menopause transition." *Recent Prog Horm Res* 57:257–75.

Burleson, M. H., W. B. Malarkey, et al. (1998). "Postmenopausal hormone replacement: Effects on autonomic, neuroendocrine, and immune reactivity to brief psychological stressors." *Psychosom Med* 60 (1): 17–25.

Buss, D. (1990). "International preferences in selecting mates: A study of 37 cultures." *Journal of CrossCultural Psychology* 21:5–47.

Buss, D. D. (2003). *Evolutionary Psychology: The New Science of Mind*, 2nd ed. New York: Allyn & Bacon.

Buss, D. M. (1989). "Conflict between the sexes: Strategic interference and the evocation of anger and upset." *J Pers Soc Psychol* 56 (5): 735–47.

Buss, D. M. (1995). "Psychological sex differences. Origins through sexual selection." *Am Psychol* 50 (3): 164–68; discussion 169–71.

Buss, D. M. (2002). "Review: Human Mate Guarding." *Neuro Endocrinol Lett* 23 (Suppl 4): 23–29.

Buss, D. M., and D. P. Schmitt (1993). "Sexual strategies theory: An evolutionary perspective on human mating." *Psychol Rev* 100 (2): 204–32.

Buster, J. E., S. A. Kingsberg, et al. (2005). "Testosterone patch for low sexual desire in surgically menopausal women: A randomized trial." *Obstet Gynecol* 105 (5, Pt. 1): 944–52.

Butler, T., H. Pan, et al. (2005). "Fear-related activity in subgenual anterior cingulate differs between men and women." *Neuroreport* 16 (11): 1233–36.

Byrnes, E. M., B. A. Rigero, et al. (2002). "Dopamine antagonists during parturition disrupt maternal care and the retention of maternal behavior in rats." *Pharmacol Biochem Behav* 73 (4): 869–75.

Cahill, L. (2003). "Sex-related influences on the neurobiology of emotionally influenced memory." *Ann NY Acad Sci* 985:163–73.

Cahill, L. (2005). "His brain, her brain." *Sci Am* 292 (5): 40–47.

Cahill, L., and A. van Stegeren (2003). "Sex-related impairment of memory for emotional events with beta-adrenergic blockade." *Neurobiol Learn Mem* 79 (1): 81–88.

Calder, A. J., A. D. Lawrence, and A. W. Young (2001). "Neuropsychology of fear and loathing." *Nature Reviews Neuroscience* 2:352–63.

Caldji, C., D. Francis, et al. (2000). "The effects of early rearing environment on the development of GABAA and central benzodiazepine receptor levels and novelty-induced fearfulness in the rat." *Neuropsychopharmacology* 22 (3): 219–29.

Call, J. D. (1998). "Extraordinary changes in behavior in an infant after a brief separation." *J Dev Behav Pediatr* 19 (6): 424–28.

Cameron, J. (2000). "Reproductive dysfunction in primates, behaviorally induced." In G. Fink, ed., *Encyclopedia of Stress*, 366–72 New York: Academic Press.

References

Cameron, J. L. (1997). "Stress and behaviorally induced reproductive dysfunction in primates." *Semin Reprod Endocrinol* 15 (1): 37–45.

Cameron, J. L. (2004). "Interrelationships between hormones, behavior, and affect during adolescence: Understanding hormonal, physical, and brain changes occurring in association with pubertal activation of the reproductive axis. Introduction to part III." *Ann NY Acad Sci* 1021:110–23.

Cameron, N. M., F. A. Champagne, et al. (2005). "The programming of individual differences in defensive responses and reproductive strategies in the rat through variations in maternal care." *Neurosci Biobehav Rev* 29 (4–5): 843–65.

Campbell, A. (1993). *Out of Control: Men, Women and Aggression.* New York: Basic Books.

Campbell, A. (1995). "A few good men: Evolutionary psychology and female adolescent aggression." *Ethology and Sociobiology* 16:99–123.

Campbell, A. (1999). "Staying alive: Evolution, culture, and women's intrasexual aggression." *Behavioral & Brain Sciences,* 22:203–14.

Campbell, A. (2002). *A Mind of Her Own: The Evolutionary Psychology of Women.* London: Oxford University Press.

Campbell, A. (2004). "Female competition: Causes, constraints, content and contexts." *J Sex Res* 41:6–26.

Campbell, A. (2005). "Aggression." In *Handbook of Evolutionary Psychology,* ed. Buss, D. 628–52 New York: Wiley. Campbell, A. L. Shirley, and J. Candy (2004). "A longitudinal study of genderrelated cognition and behavior." *Developmental Science* 7:1–9.

Camras, L. A., S. Ribordy, et al. (1990). "Maternal facial behavior and the recognition and production of emotional expression by maltreated and nonmaltreated children." *Dev Psychol* 26 (2): 304–12.

Canli, T., J. E. Desmond, et al. (2002). "Sex differences in the neural basis of emotional memories." *Proc Natl Acad Sci USA* 99 (16): 10789–94.

Cannon, W. B. (1932). *The Wisdom of the Body.* New York: W. W. Norton.

Capitanio, J. P., S. P. Mendoza, et al. (2005). "Rearing environment and hypothalamic-pituitary-adrenal regulation in young rhesus monkeys (*Macaca mulatta*)." *Dev Psychobiol* 46 (4): 318–30.

Cardinal, R. N., C. A. Winstanley, et al. (2004). "Limbic corticostriatal systems and delayed reinforcement." *Ann NY Acad Sci* 1021:33–50.

Carey, W. B., and S. C. McDevitt (1978). "Revision of the infant temperament questionnaire." *Pediatrics* 61 (5): 735–39.

Carter, C. S. (1992). "Oxytocin and sexual behavior." *Neurosci Biobehav Rev* 16 (2): 131–44.

Carter, C. S. (1998). "Neuroendocrine perspectives on social attachment and love." *Psychoneuroendocrinology* 23 (8): 779–818.

Carter, C. S. (2003). "Developmental consequences of oxytocin." *Physiol Behav* 79 (3): 383–97.

Carter, C. S. (2004). "Proximate mechanisms regulating sociality and social monogamy, in the context of evolution." In *The origin and nature of sociality*, ed. R. D. Sussman, Piscataway, NJ: Aldine Transaction.

Carter, C. S. (2006). Personal communication.

Carter, C. S., and M. Altemus (1997). "Integrative functions of lactational hormones in social behavior and stress management." *Ann NY Acad Sci* 807:164–74.

Carter, C. S., A. C. DeVries, et al. (1995). "Physiological substrates of mammalian monogamy: The prairie vole model." *Neurosci Biobehav Rev* 19 (2): 303–14.

Carter, C. S., A. C. DeVries, et al. (1997). "Peptides, steroids, and pair bonding." *Ann NY Acad Sci* 807:260–72.

Cashdan, E. (1995). "Hormones, sex, and status in women." *Horm Behav* 29 (3): 354–66.

Caspi, A., K. Sugden, et al. (2003). "Influence of life stress on depression: Moderation by a polymorphism in the 5HTT gene." *Science* 301 (5631): 386–89.

Cassidy, J. (2001). "Gender differences among newborns on a transient otoacoustic emissions test for hearing." *Journal of Music Therapy* 37:28–35.

Champagne, F., J. Diorio, et al. (2001). "Naturally occurring variations in maternal behavior in the rat are associated with differences in estrogen-inducible central oxytocin receptors." *Proc Natl Acad Sci USA* 98 (22): 12736–41.

Champagne, F., and M. J. Meaney (2001). "Like mother, like daughter: Evidence for non-genomic transmission of parental behavior and stress responsivity." *Prog Brain Res* 133:287–302.

Champagne, F. A., D. D. Francis, et al. (2003). "Variations in maternal care in the rat as a mediating influence for the effects of environment on development." *Physiol Behav* 79 (3): 359–71.

Champagne, F. A., I. C. Weaver, et al. (2003). "Natural variations in maternal care are associated with estrogen receptor alpha expression and estrogen sensitivity in the medial preoptic area." *Endocrinology* 144 (11): 4720–24.

Charmandari, E., C. Tsigos, et al. (2005). "Endocrinology of the stress response." *Annu Rev Physiol* 67:259–84.

Cherney, I. D., and M. L. Collaer (2005). "Sex differences in line judgment: Relation to mathematics preparation and strategy use." *Percept Mot Skills* 100 (3, Pt. 1): 615–27.

Chezem, J., P. Montgomery, et al. (1997). "Maternal feelings after cessation of breastfeeding: Influence of factors related to employment and duration." *J Perinat Neonatal Nurs* 11 (2): 61–70.

Chivers, M. L., G. Rieger, et al. (2004). "A sex difference in the specificity of sexual arousal." *Psychol Sci* 15 (11): 736–44.

Clarkson, T. B., and S. E. Appt (2005). "Controversies about HRT—lessons from monkey models." *Maturitas* 51 (1): 64–74.

Cohen, I. T., B. B. Sherwin, et al. (1987). "Food cravings, mood, and the menstrual cycle." *Horm Behav* 21 (4): 457–70.

Collaer, M. L., M. E. Geffner, et al. (2002). "Cognitive and behavioral characteristics of Turner syndrome: Exploring a role for ovarian hormones in female sexual differentiation." *Horm Behav* 41 (2): 139–55.

Collaer, M. L., and M. Hines (1995). "Human behavioral sex differences: A role for gonadal hormones during early development?" *Psychol Bull* 118 (1): 55–107.

Colson, M. H., A. Lemaire, et al. (2006). "Sexual behaviors and mental perception, satisfaction and expectations of sex life in men and women in France." *J Sex Med* 3 (1): 121–31.

Connell, K., M. K. Guess, et al. (2005). "Effects of age, menopause, and comorbidities on neurological function of the female genitalia." *Int J Impot Res* 17 (1): 63–70.

Connell, K., M. K. Guess, et al. (2005). "Evaluation of the role of pudendal nerve integrity in female sexual function using noninvasive techniques." *Am J Obstet Gynecol* 192 (5): 1712–17.

Connellan, J. (2000). "Sex differences in human neonatal social perception." *Infant Brain and Development* 23:113–18.

Cooke, B. (2005). "Sexually dimorphic synaptic organization of the medial amygdala." *J Neurosci* 25 (46): 10759–67.

Cooke, B. M., and C. S. Woolley (2005). "Gonadal hormone modulation of dendrites in the mammalian CNS." *J Neurobiol* 64 (1): 34–46.

Coplan, J. D., M. Altemus, et al. (2005). "Synchronized maternal-infant elevations of primate CSF CRF concentrations in response to variable foraging demand." *CNS Spectr* 10 (7): 530–36.

Corso, J. (1959). "Age and sex differences in thresholds." *Journal of the Acoustical Society of America* 31:489–507.

Cote, S., R. E. Tremblay, et al. (2002). "Childhood behavioral profiles leading to adolescent conduct disorder: Risk trajectories for boys and girls." *J Am Acad Child Adolesc Psychiatry* 41 (9): 1086–94.

Craig, I. W., E. Harper, et al. (2004). "The genetic basis for sex differences in human behaviour: Role of the sex chromosomes." *Ann Hum Genet* 68 (Pt. 3): 269–84.

Craik, F. (1977). *The Handbook of Aging and Cognition.* San Diego: Academic Press.

Crawford, J. (1992). *Emotion and Gender: Constructing Meaning from Memory.* London: Sage.

Crick, N. R., M. A. Bigbee, et al. (1996). "Gender differences in children's normative beliefs about aggression: How do I hurt thee? Let me count the ways." *Child Dev* 67 (3): 1003–14.

Cross, S. E., and L. Madson (1997). "Models of the self: Self-construals and gender." *Psychol Bull* 122 (1): 5–37.

Cummings, J. A. and L. Brizendine (2002). "Comparison of physical and emotional side effects of progesterone or medroxy-progesterone in early postmenopausal women." *Menopause* 9 (4): 253–63.

Cushing, B. S., and C. S. Carter (2000). "Peripheral pulses of oxytocin increase partner preferences in female, but not male, prairie voles." *Horm Behav* 37 (1): 49–56.

Cushing, B. S., and K. M. Kramer (2005). "Mechanisms underlying epigenetic effects of early social experience: The role of neuropeptides and steroids." *Neurosci Biobehav Rev* 29 (7): 1089–105.

Cyranowski, J. M., E. Frank, et al. (2000). "Adolescent onset of the gender difference in lifetime rates of major depression: A theoretical model." *Arch Gen Psychiatry* 57 (1): 21–27.

Dahlen, E. (2004). "Boredom proneness in anger and aggression: Effects of impulsiveness and sensation seeking." *Personality and Individual Differences* 37:1615–27.

Darnaudery, M., I. Dutriez, et al. (2004). "Stress during gestation induces lasting effects on emotional reactivity of the dam rat." *Behav Brain Res* 153 (1): 211–16.

Davidson, K. M. (1996). "Coder gender and potential for hostility ratings." *Health Psychology* 15 (4): 298–302.

Davis, S. R. (1998). "The role of androgens and the menopause in the female sexual response." *Int J Impot Res* 10 (Suppl. 2): S82–83; discussion S98–101.

Davis, S. R., I. Dinatale, et al. (2005). "Postmenopausal hormone therapy: From monkey glands to transdermal patches." *J Endocrinol* 185 (2): 207–22.

Davis, S. R., and J. Tran (2001). "Testosterone influences libido and well being in women." *Trends Endocrinol Metab* 12 (1): 33–37.

Davison, S. L., R. Bell, et al. (2005). "Androgen levels in adult females: Changes with age, menopause, and oophorectomy." *J Clin Endocrinol Metab* 90 (7): 3847–53.

Dawood, K., K. M. Kirk, et al. (2005). "Genetic and environmental influences on the frequency of orgasm in women." *Twin Res Hum Genet* 8 (1): 27–33.

de Kloet, E. R., R. M. Sibug, et al. (2005). "Stress, genes and the mechanism of programming the brain for later life." *Neurosci Biobehav Rev* 29 (2): 271–81.

de Waal, F. B. (2005). "A century of getting to know the chimpanzee." *Nature* 437 (7055): 56–59.

De Wied, D. (1997). "Neuropeptides in learning and memory process." *Behav Brain Res* 83:83–90.

Deacon, T. (1997). *The Co-Evolution of Language and the Brain.* New York: W. W. Norton.

Debiec, J. (2005). "Peptides of love and fear: Vasopressin and oxytocin modulate the integration of information in the amygdala." *Bioessays* 27 (9): 869–73.

Deckner, D. F. A. (2003). "Rhythm in mother–infant interactions." *Infancy* 4 (2): 201–17.

DeJudicibus, M. A., and M. P. McCabe (2002). "Psychological factors and the sexuality of pregnant and postpartum women." *J Sex Res* 39 (2): 94–103.

Dennerstein, L., E. C. Dudley, et al. (1997). "Sexuality, hormones and the menopausal transition." *Maturitas* 26 (2): 83–93.

Dennerstein, L., E. Dudley, et al. (1997). "Wellbeing and the menopausal transition." *J Psychosom Obstet Gynaecol* 18 (2): 95–101.

Dennerstein, L., E. Dudley, et al. (2000). "Life satisfaction, symptoms, and the menopausal transition." *Medscape Womens Health* 5 (4): E4.

Denton, D., R. Shade, et al. (1999). "Neuroimaging of genesis and satiation of thirst and an interoceptor-driven theory of origins of primary consciousness." *Proc Natl Acad Sci USA* 96 (9): 5304–9.

Depue, R., J. Morrone-Stupinsky (2005). "A neurobiobehavioral model of affiliative bonding: implications for conceptualizing a human trait of affiliation." *Behav Brain Sci* 28:313–50.

Derbyshire, S. W., T. E. Nichols, et al. (2002). "Gender differences in patterns of cerebral activation during equal experience of painful laser stimulation." *J Pain* 3 (5): 401–11.

DeRubeis, R. J., S. D. Hollon, et al. (2005). "Cognitive therapy vs medications in the treatment of moderate to severe depression." *Arch Gen Psychiatry* 62 (4): 409–16.

DeVries, A. C., M. B. DeVries, et al. (1995). "Modulation of pair bonding in female prairie voles (Microtus ochrogaster) by corticosterone." *Proc Natl Acad Sci USA* 92 (17): 7744–48.

DeVries, A. C., M. B. DeVries, et al. (1996). "The effects of stress on social preferences are sexually dimorphic in prairie voles." *Proc Natl Acad Sci USA* 93 (21): 11980–84.

DeVries, A. C., T. Gupta, et al. (2002). "Corticotropin-releasing factor induces social preferences in male prairie voles." *Psychoneuroendocrinology* 27 (6): 705–14.

DeVries, A. C., S. E. Taymans, et al. (1997). "Social modulation of corticosteroid responses in male prairie voles." *Ann NY Acad Sci* 807:494–97.

REFERENCES

DeVries, G. J. (1999). "Brain sexual dimorphism and sex differences in parental and other social behaviors." In C. S. Carter, I. I. Lederhendler, and B. Kirkpatrick, eds., *The Integrative Neurobiology of Affiliation,* 155–68. Cambridge, MA: MIT Press.

Dluzen, D. E. (2005). "Estrogen, testosterone, and gender differences." *Endocrine* 27 (3): 259–68.

Dluzen, D. E. (2005). "Unconventional effects of estrogen uncovered." *Trends Pharmacol Sci* 26 (10): 485–87.

Dobson, H., S. Ghuman, et al. (2003). "A conceptual model of the influence of stress on female reproduction." *Reproduction* 125 (2): 151–63.

Dodge, K. A., J. D. Coie, et al. (1982). "Behavior patterns of socially rejected and neglected preadolescents: The roles of social approach and aggression." *J Abnorm Child Psychol* 10 (3): 389–409.

Douda, D. (2005). *Women turning to custom hormone therapy.* WCCO TV, Kansas City December 14, 2005.

Douma, S. L., C. Husband, et al. (2005). "Estrogen-related mood disorders: Reproductive life cycle factors." *ANS Adv Nurs Sci* 28 (4): 364–75.

Dreher, J., P. Schmidt, et al. (2005). "Menstrual cycle phase modulates the reward system in women." Society for Neuroscience meeting, Washington, D.C.

Dunbar, R. (1996). *Grooming, Gossip, and the Evolution of Language.* Cambridge, MA: Harvard University Press.

Dunn, K., L. Cherkas, and T. Spector (2005). "Genes drive ability to orgasm." *Biol Letter,* 5 (2) 308.

Duval, F., M. C. Mokrani, et al. (1999). "Thyroid axis activity and serotonin function in major depressive episode." *Psychoneuroendocrinology* 24 (7): 695–712.

Eagly, A. H. (1986). "Gender and aggressive behavior: A meta-analytic review of the social psychological literature." *Psychol Bull* 100 (2): 309–30.

Eberhard, W. G. (1996). *Female Control: Sexual Selection by Cryptic Female Choice.* Princeton: Princeton University Press.

Edhborg, M., M. Friberg, et al. (2005). " 'Struggling with life': Narratives from women with signs of postpartum depression." *Scand J Public Health* 33 (4): 261–67.

Editorial (2005). "Menstruation and reproduction in the context of therapy: Required reading for all therapists." *Psychology of Women Quarterly*, 29 (3): 340–41.

Eisenberg, N. (1996). "Gender development and gender effects." In *The Handbook of Educational Psychology*, ed. D. C. Berliner. New York: Macmillan. 121–39.

Eisenberg, N., R. A. Fabes, et al. (1993). "The relations of emotionality and regulation to preschoolers' social skills and sociometric status." *Child Dev* 64 (5): 1418–38.

Eisenberg, N., R. A. Fabes, et al. (1993). "The relations of empathy-related emotions and maternal practices to children's comforting behavior." *J Exp Child Psychol* 55 (2): 131–50.

Eisenberger, N. I., and M. D. Lieberman (2004). "Why rejection hurts: A common neural alarm system for physical and social pain." *Trends Cogn Sci* 8 (7): 294–300.

Ekstrom, H. (2005). "Trends in middleaged women's reports of symptoms, use of hormone therapy and attitudes towards it." *Maturitas* 52 (2): 154–64.

Elavsky, S., E. McAuley, et al. (2005). "Physical activity enhances longterm quality of life in older adults: Efficacy, esteem, and affective influences." *Ann Behav Med* 30 (2): 138–45.

Elavsky, S., and E. McAuley (2005): "Physical activity, symptoms, esteem, and life satisfaction during menopause." *Maturitas* 52 (3–4): 374–85.

Else-Quest, N. M., J. S. Hyde, et al. (2006). "Gender differences in temperament: a meta-analysis." *Psychol Bull* 132 (1): 33–72.

Emanuele, E., P. Politi, et al. (2006). "Raised plasma nerve growth factor levels associated with early-stage romantic love." *Psychoneuroendocrinology*. In press.

Enserink, M. (2005). "Let's talk about sex—and drugs." *Science* 308 (5728): 1578.

Epel, E. S., E. H. Blackburn, et al. (2004). "Accelerated telomere shortening in response to life stress." *Proc Natl Acad Sci USA* 101 (49): 17312–15.

Epel, E., S. Jimenez, et al. (2004). "Are stress eaters at risk for the metabolic syndrome?" *Ann NY Acad Sci* 1032:208–10.

Epel, E., Jue Lin, et al. (2006). "Cell aging in relation to stress arousal

and cardiovascular disease risk factors." *Psychoneuroendocrinology.*
In press.

Erickson, K. I., S. J. Colcombe, et al. (2005). "Selective sparing of brain
tissue in postmenopausal women receiving hormone replacement
therapy." *Neurobiol Aging* 26 (8): 1205–13.

Erwin, R. J., R. C. Gur, et al. (1992). "Facial emotion discrimination: I.
Task construction and behavioral findings in normal subjects."
Psychiatry Res 42 (3): 231–40.

Esch, T., and G. B. Stefano (2005). "The neurobiology of love." *Neuro
Endocrinol Lett* 26 (3): 175–92.

Estanislau, C., and S. Morato (2005). "Prenatal stress produces more
behavioral alterations than maternal separation in the elevated
plusmaze and in the elevated T-maze." *Behav Brain Res* 163 (1):
70–77.

Eysenck, S. B., and H. J. Eysenck (1978). "Impulsiveness and venture-
someness: Their position in a dimensional system of personality
description." *Psychol Rep* 43 (3, Pt. 2): 1247–55.

Faber, R. (1994). "Physiological, emotional and behavioral correlates
of gender segregation." In *Childhood Gender Segregation: Causes and
Consequences,* ed. C. Leaper. San Francisco: JosseyBass. p. 234–302.

Fagot, B. I., R. Hagan, et al. (1985). "Differential reactions to assertive
and communicative acts of toddler boys and girls." *Child Dev* 56
(6): 1499–505.

Fagot, B. I., and M. D. Leinbach (1989). "The young child's gender
schema: Environmental input, internal organization." *Child Dev* 60
(3): 663–72.

Farr, S. A., W. A. Banks, et al. (2000). "Estradiol potentiates
acetylcholine and glutamate-mediated post-trial memory
processing in the hippocampus." *Brain Res* 864 (2): 263–69.

Farroni, T., M. Johnson, et al. (2005). "Newborns' preference for face-
relevant stimuli: Effects of contrast polarity." *Proc Natl Acad Sci
USA* 102 (47): 17245–50.

Featherstone, R. E., A. S. Fleming, et al. (2000). "Plasticity in the
maternal circuit: Effects of experience and partum condition on brain
astrocyte number in female rats." *Behav Neurosci* 114 (1): 158–72.

Feingold, A. (1994). "Gender differences in personality: A
meta-analysis." *Psychol Bull* 116 (3): 429–56.

Ferguson, J. N., J. M. Aldag, et al. (2001). "Oxytocin in the medial amygdala is essential for social recognition in the mouse." *J Neurosci* 21 (20): 8278–85.

Ferguson, T., and H. Eyre (2000). "Engendering gender differences in shame and guilt: Stereotypes, socialization and situational pressures." In *Gender and Emotion: Social Psychological Prespectives,* ed. A. H. Fisher, 254–76 Cambridge: Cambridge University Press.

Fernandez-Guasti, A., F. P. Kruijver, et al. (2000). "Sex differences in the distribution of androgen receptors in the human hypothalamus." *J Comp Neurol* 425 (3): 422–35.

Ferris, C. F., P. Kulkarni, et al. (2005). "Pup suckling is more rewarding than cocaine: Evidence from functional magnetic resonance imaging and three-dimensional computational analysis." *J Neurosci* 25 (1): 149–56.

Finch, C. (2002). "Evolution and the plasticity of aging in the reproductive schedules in long-lived animals: The importance of genetic variation in neuroendocrine mechanisms." In *Hormones, Brain and Behavior,* ed. D. W. Pfaff, vol. 4, 799–820. San Diego: Academic Press.

Fink, G., B. E. Sumner, et al. (1998). "Sex steroid control of mood, mental state and memory." *Clin Exp Pharmacol Physiol* 25 (10): 764–75.

Fischer, U., C. W. Hess, et al. (2005). "Uncrossed cortico-muscular projections in humans are abundant to facial muscles of the upper and lower face, but may differ between sexes." *J Neurol* 252 (1): 21–26.

Fish, E. W., D. Shahrokh, et al. (2004). "Epigenetic programming of stress responses through variations in maternal care." *Ann NY Acad Sci* 1036:167–80.

Fisher, H. (2004). *Why We Love: The Nature and Chemistry of Romantic Love.* New York: Henry Holt.

Fisher, H. (2005). Personal communication.

Fisher, H., A. Aron, et al. (2005). "Romantic love: An fMRI study of a neural mechanism for mate choice." *J Comp Neurol* 493 (1): 58–62.

Fisher, H. E., A. Aron, et al. (2002). "Defining the brain systems of lust, romantic attraction, and attachment." *Arch Sex Behav* 31 (5): 413–19.

REFERENCES

Fivush, R., and N. R. Hamond (1989). "Time and again: Effects of repetition and retention interval on 2 year olds' event recall." *J Exp Child Psychol* 47 (2): 259–73.

Flannery, K. A., and M. W. Watson (1993). "Are individual differences in fantasy play related to peer acceptance levels?" *J Genet Psychol* 154 (3): 407–16.

Fleming, A. S., C. Corter, et al. (1993). "Postpartum factors related to mother's attraction to newborn infant odors." *Dev Psychobiol* 26 (2): 115–32.

Fleming, A. S., C. Corter, et al. (2002). "Testosterone and prolactin are associated with emotional responses to infant cries in new fathers." *Horm Behav* 42 (4): 399–413.

Fleming, A. S., E. Klein, et al. (1992). "The effects of a social support group on depression, maternal attitudes and behavior in new mothers." *J Child Psychol Psychiatry* 33 (4): 685–98.

Fleming, A. S., G. W. Kraemer, et al. (2002). "Mothering begets mothering: The transmission of behavior and its neurobiology across generations." *Pharmacol Biochem Behav* 73 (1): 61–75.

Fleming, A. S., D. H. O'Day, et al. (1999). "Neurobiology of mother-infant interactions: Experience and central nervous system plasticity across development and generations." *Neurosci Biobehav Rev* 23 (5): 673–85.

Fleming, A. S., D. Ruble, et al. (1997). "Hormonal and experiential correlates of maternal responsiveness during pregnancy and the puerperium in human mothers." *Horm Behav* 31 (2): 145–58.

Fleming, A. S., and J. Sarker (1990). "Experience-hormone interactions and maternal behavior in rats." *Physiol Behav* 47 (6): 1165–73.

Fleming, A. S., M. Steiner, et al. (1997). "Cortisol, hedonics, and maternal responsiveness in human mothers." *Horm Behav* 32 (2): 85–98.

Forger, N. G., G. J. Rosen, et al. (2004). "Deletion of Bax eliminates sex differences in the mouse forebrain." *Proc Natl Acad Sci USA* 101 (37): 13666–71.

Forger, N. G. (2006). "Cell death and sexual differentiation of the nervous system." *Neuroscience* 138 (3): 929–38.

Fox, C., H. S. Wolff, and J. A. Baker (1970). "Measurement of intra-vaginal and intrauterine pressures human coitus by radiotelemetry." *J Reprod Fert* 22:243–51.

Francis, D., J. Diorio, et al. (1999). "Nongenomic transmission across generations of maternal behavior and stress responses in the rat." *Science* 286 (5442): 1155–58.

Francis, D. D., F. A. Champagne, et al. (1999). "Maternal care, gene expression, and the development of individual differences in stress reactivity." *Ann NY Acad Sci* 896:66–84.

Francis, D. D., J. Diorio, et al. (2002). "Environmental enrichment reverses the effects of maternal separation on stress reactivity." *J Neurosci* 22 (18): 7840–43.

Francis, D. D., and M. J. Meaney (1999). "Maternal care and the development of stress responses." *Curr Opin Neurobiol* 9 (1): 128–34.

Francis, D. D., L. J. Young, et al. (2002). "Naturally occurring differences in maternal care are associated with the expression of oxytocin and vasopressin (V1a) receptors: Gender differences." *J Neuroendocrinol* 14 (5): 349–53.

Franklin, T. (2006). "Sex and ovarian steroids modulate brain-derived neurotrophic factor (BDNF) protein levels in rat hippocampus under stressful and non-stressful conditions." *Psychoneuroendocrinology* 31 1:38–48.

Freeman, E. W. (2004). "Luteal phase administration of agents for the treatment of premenstrual dysphoric disorder." *CNS Drugs* 18 (7): 453–68.

Frey, W. (1985). "Crying: The mystery of tears." *Winston Pr* (September, 1985).

Fries, A. B., T. E. Ziegler, et al. (2005). "Early experience in humans is associated with changes in neuropeptides critical for regulating social behavior." *Proc Natl Acad Sci USA* 102 (47): 17237–40.

Frodi, A. (1977). "Sex differences in perception of a provocation, a survey." *Percept Mot Skills* 44 (1): 113–14.

Frodi, A., J. Macaulay, et al. (1977). "Are women always less aggressive than men? A review of the experimental literature." *Psychol Bull* 84 (4): 634–60.

Fry, D. P. (1992). "Female aggression among the Zapotec of Oaxaca, Mexico." In K. Bjorkqvist and P. Niemela, eds., *Of Mice and Women: Aspects of Female Aggression,* 187–200. San Diego: Academic Press.

REFERENCES

Fujita, F., E. Diener, et al. (1991). "Gender differences in negative affect and well-being: The case for emotional intensity." *J Pers Soc Psychol* 61 (3): 427–34.

Furuta, M., and R. S. Bridges (2005). "Gestation-induced cell proliferation in the rat brain." *Brain Res Dev Brain Res* 156 (1): 61–66.

Gangestad, S. W., and R. Thornhill (1998). "Menstrual cycle variation in women's preferences for the scent of symmetrical men." *Proc Biol Sci* 265 (1399): 927–33.

Garner, A. (1997). *Conversationally Speaking.* New York: McGraw-Hill.

Garstein, M. (2003). "Studying infant temperament." Infant Behavior *and Development,* 26:64–86.

Gatewood, J. D., and M. D. Morgan, et al. (2005). "Motherhood mitigates aging-related decrements in learning and memory and positively affects brain aging in the rat." *Brain Res Bull* 66 (2): 91–98.

Genazzani, A. D. (2005). "Neuroendocrine aspects of amenorrhea related to stress." *Pediatr Endocrinol Rev* 2 (4): 661–68.

Getchell, T. (1991). *Smell and Taste in Health and Disease.* New York: Raven Press.

Giammanco, M., G. Tabacchi, et al. (2005). "Testosterone and aggressiveness." *Med Sci Monit* 11 (4): RA 136–45.

Giedd, J. (2005). Personal communication.

Giedd, J. N. (2003). "The anatomy of mentalization: A view from developmental neuroimaging." *Bull Menninger Clin* 67 (2): 132–42.

Giedd, J. N. (2004). "Structural magnetic resonance imaging of the adolescent brain." *Ann NY Acad Sci* 1021:77–85.

Giedd, J. N., J. Blumenthal, et al. (1999). "Brain development during childhood and adolescence: A longitudinal MRI study." *Nat Neurosci* 2 (10): 861–63.

Giedd, J. N., F. X. Castellanos, et al. (1997). "Sexual dimorphism of the developing human brain." *Prog Neuropsychopharmacol Biol Psychiatry* 21 (8): 1185–201.

Giedd, J. N., J. M. Rumsey, et al. (1996). "A quantitative MRI study of the corpus callosum in children and adolescents." *Brain Res Dev Brain Res* 91 (2): 274–80.

Giedd, J. N., J. W. Snell, et al. (1996). "Quantitative magnetic resonance imaging of human brain development: Ages 4–18." *Cereb Cortex* 6 (4): 551–60.

Giedd, J. N., A. C. Vaituzis, et al. (1996). "Quantitative MRI of the temporal lobe, amygdala, and hippocampus in normal human development: Ages 4–18 years." *J Comp Neurol* 366 (2): 223–30.

Giltay, E. J., K. H. Kho, et al. (2005). "The sex difference of plasma homovanillic acid is unaffected by cross-sex hormone administration in transsexual subjects." *J Endocrinol* 187 (1): 109–16.

Gingrich, B., Y. Liu, et al. (2000). "Dopamine D2 receptors in the nucleus accumbens are important for social attachment in female prairie voles (*Microtus ochrogaster*)." *Behav Neurosci* 114 (1): 173–83.

Gizewski, E. R., E. Krause, et al. (2006). "Gender-specific cerebral activation during cognitive tasks using functional MRI: Comparison of women in mid-luteal phase and men." *Neuroradiology* 48 (1): 14–20.

Glazer, I. M. (1992). "Interfemale aggression and resource scarcity in a crosscultural perspective." In K. Bjorkqvist and P. Niemela, eds., *Of Mice and Women: Aspects of Female Aggression*, 163–72. San Diego: Academic Press.

Glickman, S. E., R. V. Short, et al. (2005). "Sexual differentiation in three unconventional mammals: Spotted hyenas, elephants and tammar wallabies." *Horm Behav* 48 (4): 403–17.

Goldstat, R., E. Briganti, et al. (2003). "Transdermal testosterone therapy improves well-being, mood, and sexual function in pre-menopausal women." *Menopause* 10 (5): 390–98.

Goldberg E., K. Podell, et al. (1994). "Cognitive bias, functional cortical geometry and the frontal lobes: laterality, sex and handed-ness." *J Cog Neurosci* 6: 276–96.

Goldstein, J. M., M. Jerram, et al. (2005). "Hormonal cycle modulates arousal circuitry in women using functional magnetic resonance imaging." *J Neurosci* 25 (40): 9309–16.

Goldstein, J. M., M. Jerram, et al. (2005). "Sex differences in prefrontal cortical brain activity during FMRI of auditory verbal working memory." *Neuropsychology* 19 (4): 509–19.

Goldstein, J. M., L. J. Seidman, et al. (2001). "Normal sexual dimorphism of the adult human brain assessed by in vivo magnetic resonance imaging." *Cereb Cortex* 11 (6): 490–97.

Golombok, S., and S. Fivush (1994). *Gender Development*. New York: Cambridge University Press.

Good, C. D., K. Lawrence, et al. (2003). "Dosage-sensitive X-linked locus influences the development of amygdala and orbitofrontal cortex, and fear recognition in humans." *Brain* 126 (Pt. 11): 2431–46.

Goos, L. M. and S. Irwin (2002). "Sex related factors in the perception of threatening facial expressions." *Journal of Nonverbal Behavior* 26 (1): 27–41.

Gootjes, L., A. Bouma, et al. (2006). "Attention modulates hemispheric differences in functional connectivity: Evidence from MEG recordings." *Neuroimage*. In press.

Goy, R. W., F. B. Bercovitch, et al. (1988). "Behavioral masculinization is independent of genital masculinization in prenatally androgenized female rhesus macaques." *Horm Behav* 22 (4): 552–71.

Graham, C. A., E. Janssen, et al. (2000). "Effects of fragrance on female sexual arousal and mood across the menstrual cycle." *Psychophysiology* 37 (1): 76–84.

Grammer, K. (1993). "Androstadienone – a male pheromone?" *Ethol Sociobiol* 14:201–7.

Gray, A., H. A. Feldman, et al. (1991). "Age, disease, and changing sex hormone levels in middle-aged men: Results of the Massachusetts Male Aging Study." *J Clin Endocrinol Metab* 73 (5): 1016–25.

Gray, P. B., B. C. Campbell, et al. (2004). "Social variables predict between-subject but not day-to-day variation in the testosterone of U.S. men." *Psychoneuroendocrinology* 29 (9): 1153–62.

Green, R. (2002). "Sexual identity and sexual orientation." In *Hormones, Brain and Behavior*, ed. D. W. Pfaff, vol. 4, 463–86. San Diego: Academic Press.

Grewen, K. M., S. S. Girdler, et al. (2005). "Effects of partner support on resting oxytocin, cortisol, norephinephrine, and blood pressure before and after warm partner contact." *Psychosom Med* 67 (4): 531–38.

Griffin, L. D. and S. H. Mellon (1999). "Selective serotonin reuptake inhibitors directly alter activity of neurosteroidogenic enzymes." *Proc Natl Acad Sci USA* 96 (23): 13512–17.

Grossman, M., and W. Wood (1993). "Sex differences in intensity of emotional experience: A social role interpretation." *J Pers Soc Psychol* 65 (5): 1010–22.

Grumbach, M. (2003). "Puberty." In *Williams Textbook of Endocrinology*, ed. R. H. Williams 1115–286. New York: W. B. Saunders Co.

Grumbach, M. (2005). Personal communication.

Grumbach, M. M. (2002). "The neuroendocrinology of human puberty revisited." *Horm Res* 57 (Suppl. 2): 2–14.

Guay, A. (2005). "Commentary on androgen deficiency in women and the FDA advisory board's recent decision to request more safety data." *Int J Impot Res* 17 (4): 375–76.

Guay, A., and S. R. Davis (2002). "Testosterone insufficiency in women: Fact or fiction?" *World J Urol* 20 (2): 106–10.

Guay, A., J. Jacobson, et al. (2004). "Serum androgen levels in healthy premenopausal women with and without sexual dysfunction: Part B: Reduced serum androgen levels in healthy premenopausal women with complaints of sexual dysfunction." *Int J Impot Res* 16 (2): 121–29.

Guay, A., and R. Munarriz, et al. (2004). "Serum androgen levels in healthy premenopausal women with and without sexual dysfunction: Part A. Serum androgen levels in women aged 20–49 years with no complaints of sexual dysfunction." *Int J Impot Res* 16 (2): 112–20.

Guay, A. T. (2002). "Screening for androgen deficiency in women: Methodological and interpretive issues." *Fertil Steril* 77 (Suppl. 4): S83–88.

Guay, A. T., and J. Jacobson (2002). "Decreased free testosterone and dehydro-epiandrosterone-sulfate (DHEAS) levels in women with decreased libido." *J Sex Marital Ther* 28 (Suppl. 1): 129–42.

Gulati, M. (2005). "Exercise may ward off death in women with metabolic syndrome." American Heart Association Scientific Sessions, Philadelphia,

Gulati, M., H. R. Black, et al. (2005). "The prognostic value of a nomogram for exercise capacity in women." *N Engl J Med* 353 (5): 468–75.

Gulinello, M., D. Lebesgue, et al. (2006). "Acute and chronic estradiol treatments reduce memory deficits induced by transient global ischemia in female rats." *Horm Behav* 49 (2): 246–60.

Gur, R. C., F. GunningDixon, et al. (2002). "Sex differences in temporo-limbic and frontal brain volumes of healthy adults." *Cereb Cortex* 12 (9): 998–1003.

Gur, R. C., F. M. GunningDixon, et al. (2002). "Brain region and sex differences in age association with brain volume: A quantitative MRI study of healthy young adults." *Am J Geriatr Psychiatry* 10 (1): 72–80.

Gur, R. C., L. H. Mozley, et al. (1995). "Sex differences in regional cerebral glucose metabolism during a resting state." *Science* 267 (5197): 528–31.

Gurung, R. A., S. E. Taylor, et al. (2003). "Accounting for changes in social support among married older adults: Insights from the MacArthur Studies of Successful Aging." *Psychol Aging* 18 (3): 487–96.

Gust, D. A., M. E. Wilson, et al. (2000). "Activity of the hypothalamic-pituitary-adrenal axis is altered by aging and exposure to social stress in female rhesus monkeys." *J Clin Endocrinol Metab* 85 (7): 2556–63.

Guthrie, J. R., L. Dennerstein, et al. (2003). "Central abdominal fat and endogenous hormones during the menopausal transition." *Fertil Steril* 79 (6): 1335–40.

Guthrie, J. R., L. Dennerstein, et al. (2003). "Health care-seeking for menopausal problems." *Climacteric* 6 (2): 112–17.

Guthrie, J. R., L. Dennerstein, et al. (2004). "The menopausal transition: A 9-year prospective population-based study: The Melbourne Women's Midlife Health Project." *Climacteric* 7 (4): 375–89.

Gutteling, B. M., C. de Weerth, et al. (2005). "The effects of prenatal stress on temperament and problem behavior of 27-month-old toddlers." *Eur Child Adolesc Psychiatry* 14 (1): 41–51.

Gutteling, B. M., C. de Weerth, et al. (2005). "Prenatal stress and children's cortisol reaction to the first day of school." *Psychoneuroendocrinology* 30 (6): 541–49.

Haier, R. J., R. E. Jung, et al. (2005). "The neuroanatomy of general intelligence: Sex matters." *Neuroimage* 25 (1): 320–27.

Halari, R., M. Hines, et al. (2005). "Sex differences and individual differences in cognitive performance and their relationship to endogenous gonadal hormones and gonadotropins." *Behav Neurosci* 119 (1): 104–17.

Halari, R., and V. Kumari (2005). "Comparable cortical activation with inferior performance in women during a novel cognitive inhibition task." *Behav Brain Res* 158 (1): 167–73.

Halari, R., V. Kumari, et al. (2004). "The relationship of sex hormones and cortisol with cognitive functioning in schizophrenia." *J Psychopharmacol* 18 (3): 366–74.

Halari, R., T. Sharma, et al. (2006). "Comparable fMRI activity with differential behavioural performance on mental rotation and overt verbal fluency tasks in healthy men and women." *Exp Brain Res* 169 (1): 1–14.

Halbreich, U. (2006). "Major depression is not a diagnosis, it is a departure point to differential diagnosis—clinical and hormonal considerations." *Psychoneuroendocrinology* 31 (1): 16–22.

Halbreich, U., L. A. Lumley, et al. (1995). "Possible acceleration of age effects on cognition following menopause." *J Psychiatr Res* 29 (3): 153–63.

Hall, J. A. (1978). "Gender effects in decoding nonverbal cues." *Psychol Bull* 85: 8845–57.

Hall, J. A. (1984). *Nonverbal sex differences: Communication accuracy and expressive style.* Baltimore: Johns Hopkins University Press.

Hall, J. A., J. D. Carter, and T. G. Horgan (2000). "Gender differences in the nonverbal communication of emotion." In A. H. Fischer, ed., *Gender and Emotion: Social Psychological Perspectives,* 97–117. London: Cambridge University Press.

Hall, L. A., A. R. Peden, et al. (2004). "Parental bonding: A key factor for mental health of college women." *Issues Ment Health Nurs* 25 (3): 277–91.

Halpern, C. T., B. Campbell, et al. (2002). "Associations between stress reactivity and sexual and nonsexual risk taking in young adult human males." *Horm Behav* 42 (4): 387–98.

Halpern, C. T., J. R. Udry, et al. (1997). "Testosterone predicts initiation of coitus in adolescent females." *Psychosom Med* 59 (2): 161–71.

Hamann, S. (2005). "Sex differences in the responses of the human amygdala." *Neuroscientist* 11 (4): 288–93.

Hamilton, W. L., M. C. Diamond, et al. (1977). "Effects of pregnancy and differential environments on rat cerebral cortical depth." *Behav Biol* 19 (3): 333–40.

Hammock, E. A., M. M. Lim, et al. (2005). "Association of vasopressin 1a receptor levels with a regulatory microsatellite and behavior." *Genes Brain Behav* 4 (5): 289–301.

Hammock, E. A., and L. J. Young (2005). "Microsatellite instability generates diversity in brain and sociobehavioral traits." *Science* 308 (5728): 1630–34.

Harman, S. M., E. A. Brinton, et al. (2004). "Is the WHI relevant to HRT started in the perimenopause?" *Endocrine* 24 (3): 195–202.

Harman, S. M., E. A. Brinton, et al. (2005). "KEEPS: The Kronos Early Estrogen Prevention Study." *Climacteric* 8 (1): 3–12.

Harman, S. M., F. Naftolin, et al. (2005). "Is the estrogen controversy over? Deconstructing the Women's Health Initiative Study: A critical evaluation of the evidence." *Ann NY Acad Sci* 1052:43–56.

Harris, G. (2004). "Pfizer gives up testing viagra on women." *New York Times*, February 28.

Harrison, K., ed. (1999). "Tales from the screen: Enduring fright reactions to scary movies." *Media Psychology*, Spring: 15–22.

Haselton, M. G., D. M. Buss, et al. (2005). "Sex, lies, and strategic interference: The psychology of deception between the sexes." *Pers Soc Psychol Bull* 31 (1): 3–23.

Hasser, C., L. Brizendine et. al. (2006). "To treat or not to treat? Depression in pregnancy and the use of SSRIs." *Current Psychiatry.* 5 (4): 31–40.

Havlicek, J. (2005). "Women prefer more dominant men for short-term mating before ovulation." *Biol Letter*, 5 (2) 217–228.

Hawkes, K. (2003). "Grandmothers and the evolution of human longevity." *Am J Hum Biol* 15 (3): 380–400.

Hawkes, K. (2004). "Human longevity: The grandmother effect." *Nature* 428 (6979): 128–29.

Hawkes, K., J. F. O'Connell, et al. (1998). "Grandmothering, menopause, and the evolution of human life histories." *Proc Natl Acad Sci USA* 95 (3): 1336–39.

Hayward, C., and K. Sanborn (2002). "Puberty and the emergence of gender differences in psychopathology." *J Adolesc Health* 30 (4 Suppl.): 49–58.

Heinrichs, M., T. Baumgartner, et al. (2003). "Social support and oxytocin interact to suppress cortisol and subjective responses to psychosocial stress." *Biol Psychiatry* 54 (12): 1389–98.

Heinrichs, M., G. Meinlschmidt, et al. (2001). "Effects of suckling on hypothalamic-pituitary-adrenal axis responses to psychosocial stress in postpartum lactating women." *J Clin Endocrinol Metab* 86 (10): 4798–804.

Heinrichs, M., I. Neumann, et al. (2002). "Lactation and stress: Protective effects of breast-feeding in humans." *Stress* 5 (3): 195–203.

Helson, R., and B. Roberts (1992). "The personality of young adult couples and wives' work patterns." *J Pers* 60 (3): 575–97.

Helson, R., and C. J. Soto (2005). "Up and down in middle age: Monotonic and nonmonotonic changes in roles, status, and personality." *J Pers Soc Psychol* 89 (2): 194–204.

Helson, R., and S. Srivastava (2001). "Three paths of adult development: Conservers, seekers, and achievers." *J Pers Soc Psychol* 80 (6): 995–1010.

Henderson, V., (2002). "Protective effects of estrogen on aging and damaged neural systems." in *Hormones, Brain and Behavior*, ed. D. W. Pfaff, vol. 4, 821–40. San Diego: Academic Press.

Henderson, V. W., J. R. Guthrie, et al. (2003). "Estrogen exposures and memory at midlife: A population-based study of women." *Neurology* 60 (8): 1369–71.

Herba, C. P., (2004). "Annotation: Development of facial expression recognition from childhood to adolescence: Behavioural and neuro-logical perspectives." *J Child Psychol Psychiatry* 45 (7): 1185–98.

Herbert, M. R., D. A. Ziegler, et al. (2005). "Brain asymmetries in autism and developmental language disorder: A nested whole-brain analysis." *Brain* 128 (1): 213–26.

Herrera, E., N. Reissland, et al. (2004). "Maternal touch and maternal child-directed speech: Effects of depressed mood in the postnatal period." *J Affect Disord* 81 (1): 29–39.

Hershberger, S. L., and N. L. Segal (2004). "The cognitive, behavioral,

References

and personality profiles of a male monozygotic triplet set discordant for sexual orientation." *Arch Sex Behav* 33 (5): 497–514.

Hickey, M., S. R. Davis, et al. (2005). "Treatment of menopausal symptoms: What shall we do now?" *Lancet* 366 (9483): 409–21.

Hill, C. A. (2002). "Gender, relationship stage, and sexual behavior: The importance of partner emotional investment within specific situations." *J Sex Res* 39 (3): 228–40.

Hill, H., F. Ott, et al. (2006). "Response execution in lexical decision tasks obscures sex-specific lateralization effects in language processing: Evidence from event-related potential measures during word reading." *Cereb Cortex*. In press.

Hill, K. (1988). "Trade offs in male and female reproductive strategies among the Ache." In *Human Reproductive Behavior: A Darwinian Perspective*, ed. Bertzig, and Borgerhoff, et al. New York: Cambridge University Press. 215–39.

Hines, M. (2002). "Sexual differentiation of human brain and behavior." In *Hormones, Brain and Behavior*, ed. D. W. Pfaff, vol. 4, 425–62. San Diego: Academic Press.

Hines, M., S. F. Ahmed, et al. (2003). "Psychological outcomes and gender-related development in complete androgen insensitivity syndrome." *Arch Sex Behav* 32 (2): 93–101.

Hines, M., C. Brook, et al. (2004). "Androgen and psychosexual development: Core gender identity, sexual orientation and recalled childhood gender role behavior in women and men with congenital adrenal hyperplasia (CAH)." *J Sex Res* 41 (1): 75–81.

Hines, M., and F. R. Kaufman (1994). "Androgen and the development of human sex-typical behavior: Rough-and-tumble play and sex of preferred playmates in children with congenital adrenal hyperplasia (CAH)." *Child Dev* 65 (4): 1042–53.

Hittelman, J. H. (1979). "Sex differences in neonatal eye contact time." *MerrillPalmer Q* 25:171–84.

Hodes, G. E., and T. J. Shors (2005). "Distinctive stress effects on learning during puberty." *Horm Behav* 48 (2): 163–71.

Holdcroft, A., L. Hall, et al. (2005). "Phosphorus-31 brain MR spectroscopy in women during and after pregnancy compared with nonpregnant control subjects." *AJNR Am J Neuroradiol* 26 (2): 352–56.

Holden, C. (2005). "Sex and the suffering brain." *Science* 308 (5728): 1574.

Holmstrom, R. (1992). "Female aggression among the great apes." In K. Bjorkqvist and P. Niemela, eds., *Of Mice and Women: Aspects of Female Aggression,* 295–306. San Diego: Academic Press.

Holstege, G., et al. (2003). "Brain activation during female sexual orgasm." *Soc Neurosci Abstr* 727:7.

Hoover-Dempsey, K. W. (1986). "Tears and weeping among professional women: In search of new understanding." *Psychology of Women Quarterly* 10:19–34.

Horgan, T. G. et al. (2004). "Gender differences in memory for the appearance of others." *Pers Soc Psychol Bull* 30 (2): 185–96.

Howard, J. M. (2002). " 'Mitochondrial Eve,' 'Y Chromosome Adam,' testosterone, and human evolution." *Riv Biol* 95 (2): 319–25.

Howes, C. (1988). "Peer interactions of young children." *Monographs of the Society for Research in Child Development,* serial no. 217, 53 (1).

Hrdy, S. (1999). *Mother Nature.* New York: Pantheon.

Hrdy, S. (2005). Personal communication.

Hrdy, S. B. (1974). "Male-male competition and infanticide among the langurs (*Presbytis entellus*) of Abu, Rajasthan." *Folia Primatol* (Basel) 22 (1): 19–58.

Hrdy, S. B. (1977). "Infanticide as a primate reproductive strategy." *Am Sci* 65 (1): 40–49.

Hrdy, S. B. (1997). "Raising Darwin's consciousness: Female sexuality and the prehominid origins of patriarchy." *Human Nature* 8 (1): 1–49.

Hrdy, S. B. (2000). "The optimal number of fathers: Evolution, demography, and history in the shaping of female mate preferences." *Ann NY Acad Sci* 907:75–96.

Huber, D., P. Veinante, et al. (2005). "Vasopressin and oxytocin excite distinct neuronal populations in the central amygdala." *Science* 308 (5719): 245–48.

Hultcrantz, M. (2006). "Estrogen and hearing: A summary of recent investigations." *Acta Otolaryngol* 126 (1): 10–14.

Hummel, T., F. Krone, et al. (2005). "Androstadienone odor thresholds in adolescents." *Horm Behav* 47 (3): 306–10.

Huot, R. L., P. A. Brennan, et al. (2004). "Negative affect in offspring of depressed mothers is predicted by infant cortisol levels at 6 months and maternal depression during pregnancy, but not postpartum." *Ann NY Acad Sci* 1032:234–36.

Hyde, J. S. (1984). "How large are gender differences in aggression? A developmental meta-analysis." *Dev Psychol* 20:722–36.

Hyde, J. S. (1988). "Gender differences in verbal ability: A meta-analysis." *Psychol Bull* 104 (1): 53–69.

Idiaka, T. (2001). "fMRI study of agerelated differences in the medial temporal lobe responses to emotional faces." Society for Neuroscience, New Orleans.

Iervolino, A. C., M. Hines, et al. (2005). "Genetic and environmental influences on sex-typed behavior during the preschool years." *Child Dev* 76 (4): 826–40.

Imperato-McGinley, J. (2002). "Gender and behavior in subjects with genetic defects in male sexual differentiation." In *Hormones, Brain and Behavior*, ed. D. W. Pfaff, vol. 5, 303–46. San Diego: Academic Press.

Insel, T. R. (2003). "Is social attachment an addictive disorder?" *Physiol Behav* 79 (3): 351–57.

Insel, T. R., and R. D. Fernald (2004). "How the brain processes social information: Searching for the social brain." *Annu Rev Neurosci* 27:697–722.

Insel, T. R., B. S. Gingrich, et al. (2001). "Oxytocin: Who needs it?" *Prog Brain Res* 133: 59–66.

Insel, T. R., and L. J. Young (2000). "Neuropeptides and the evolution of social behavior." *Curr Opin Neurobiol* 10 (6): 784–89.

Institute of Medicine. (2003). *Gender issues in medicine: Working-Group on Gender Issues in Medicine*. Institute of Medicine, November.

Irwing, P., and R. Lynn (2005). "Sex differences in means and variability on the progressive matrices in university students: A meta-analysis." *Br J Psychol* 96 (Pt. 4): 505–24.

Jacklin, C., and E. Maccoby (1978). "Social behavior at thirty-three months in same-sex and mixed-sex dyads." *Child Dev* 49: 557–69.

Jackson, A., D. Stephens, et al. (2005). "Gender differences in response to lorazepam in a human drug discrimination study." *J Psychopharmacol* 19 (6): 614–19.

Jasnow, A. M., J. Schulkin, et al. (2006). "Estrogen facilitates fear conditioning and increases corticotropin-releasing hormone mRNA expression in the central amygdala in female mice." *Horm Behav* 49 (2): 197–205.

Jausovec, N., and K. Jausovec (2005). "Sex differences in brain activity related to general and emotional intelligence." *Brain Cogn* 59 (3): 277–86.

Jawor, J. M., R. Young, et al. (2006). "Females competing to reproduce: Dominance matters but testosterone may not." *Horm Behav* 49 (3): 362–68.

Jenkins, W. J., and J. B. Becker (2003). "Dynamic increases in dopamine during paced copulation in the female rat." *Eur J Neurosci* 18 (7): 1997–2001.

Jensvold, M. E. (1996). *Psychopharmacology and women: Sex, gender and hormones.* Washington: APA Press.

Joffe, H., and L. S. Cohen (1998). "Estrogen, serotonin, and mood disturbance: Where is the therapeutic bridge?" *Biol Psychiatry* 44 (9): 798–811.

Joffe, H., L. S. Cohen, et al. (2003). "Impact of oral contraceptive pill use on premenstrual mood: Predictors of improvement and deterioration." *Am J Obstet Gynecol* 189 (6): 1523–30.

Joffe, H., J. E. Hall, et al. (2002). "Vasomotor symptoms are associated with depression in perimenopausal women seeking primary care." *Menopause* 9 (6): 392–98.

Joffe, H., C. N. Soares, et al. (2003). "Assessment and treatment of hot flushes and menopausal mood disturbance." *Psychiatr Clin North Am* 26 (3): 563–80.

Joffe, H. (2006). Personal communication.

Johns, J. M., D. A. Lubin, et al. (2004). "Gestational treatment with cocaine and fluoxetine alters oxytocin receptor number and binding affinity in lactating rat dams." *Int J Dev Neurosci* 22 (5–6): 321–28.

Johnston, A. L., and S. E. File (1991). "Sex differences in animal tests of anxiety." *Physiol Behav* 49 (2): 245–50.

Jones, B. A., and N. V. Watson (2005). "Spatial memory performance in androgen insensitive male rats." *Physiol Behav* 85 (2): 135–41.

Jones, N. A., T. Field, et al. (2004). "Greater right frontal EEG asymmetry and nonemphathic behavior are observed in children prenatally exposed to cocaine." Int J Neurosci 114 (4): 459–80.

Jordan, K., T. Wustenberg, et al. (2002). "Women and men exhibit different cortical activation patterns during mental rotation tasks." *Neuropsychologia* 40 (13): 2397–408.

Jorm, A. F., K. B. Dear, et al. (2003). "Cohort difference in sexual orientation: Results from a large age-stratified population sample." *Gerontology* 49 (6): 392–95.

Josephs, R. A., H. R. Markus, et al. (1992). "Gender and self-esteem." *J Pers Soc Psychol* 63 (3): 391–402.

Jovanovic, T., S. Szilagyi, et al. (2004). "Menstrual cycle phase effects on prepulse inhibition of acoustic startle." *Psychophysiology* 41 (3): 401–6.

Kaiser, J. (2005). "Gender in the pharmacy: Does it matter?" *Science* 308 (5728): 1572.

Kaiser, S., and N. Sachser (2005). "The effects of prenatal social stress on behaviour: Mechanisms and function." *Neurosci Biobehav Rev* 29 (2): 283–94.

Kajantie, E. (2006). "The effects of sex and hormonal status on the physiological response to acute psychosocial stress." *Psychoneuroendocrinology* 31 (2): 151–78.

Kanin, E. (1970). "A research note on male-female differentials in the experience of heterosexual love." *J Sex Res* 6 (1): 64–72.

Kaufman, J., B. Z. Yang, et al. (2004). "Social supports and serotonin transporter gene moderate depression in maltreated children." *Proc Natl Acad Sci USA* 101 (49): 17316–21.

Kaufman, J. M., and A. Vermeulen (2005). "The decline of androgen levels in elderly men and its clinical and therapeutic implications." *Endocr Rev* 26 (6): 833–76.

Keller-Wood, M., J. Silbiger, et al. (1988). "Progesterone attenuates the inhibition of adrenocorticotropin responses by cortisol in nonpregnant ewes." *Endocrinology* 123 (1): 647–51.

Kendler, K. S., M. Gatz, et al. (2006). "A Swedish national twin study of lifetime major depression." Am J Psychiatry 163 (1): 109–14.

Kendler, K. S., L. M. Thornton, et al. (2000). "Stressful life events and previous episodes in the etiology of major depression in women: An evaluation of the 'kindling' hypothesis." *Am J Psychiatry* 157 (8): 1243–51.

Kendrick, K. M. (2000). "Oxytocin, motherhood and bonding." *Exp Physiol* 85 (Spec. No.): 111S–124S.

Kendrick, K. M., A. P. Da Costa, et al. (1997). "Neural control of maternal behavior and olfactory recognition of offspring." *Brain Res Bull* 44:383–95.

Kendrick, K. M., F. Levy, et al. (1992). "Changes in the sensory processing of olfactory signals induced by birth in sleep." *Science* 256 (5058): 833–36.

Kenyon, C. (2005). Personal communication.

Kenyon, C. (2005). "The plasticity of aging: Insights from long-lived mutants." *Cell* 120 (4): 449–60.

Keverne, E. B., C. M. Nevison, and F. L. Martel (1999). "Early learning and the social bond." In C. S. Carter, I. I. Lederhendler, and B. Kirkpatrick, eds., *The Integrative Neurobiology of Affiliation*, 263–74. Cambridge, MA: MIT Press.

Kiecolt-Glaser, J. K., R. Glaser, et al. (1998). "Marital stress: Immunologic, neuroendocrine, and autonomic correlates." *Ann NY Acad Sci* 840:656–63.

Kiecolt-Glaser, J. K., T. J. Loving, et al. (2005). "Hostile marital interactions, proinflammatory cytokine production, and wound healing." *Arch Gen Psychiatry* 62 (12): 1377–84.

Kiecolt-Glaser, J. K., T. Newton, et al. (1996). "Marital conflict and endocrine function: Are men really more physiologically affected than women?" *J Consult Clin Psychol* 64 (2): 324–32.

Kimura, K., M. Ote, et al. (2005). "Fruitless specifies sexually dimorphic neural circuitry in the Drosophila brain." *Nature* 438 (7065): 229–33.

Kinnunen, A. K., J. I. Koenig, et al. (2003). "Repeated variable prenatal stress alters pre- and postsynaptic gene expression in the rat frontal pole." *J Neurochem* 86 (3): 736–48.

Kinsley, C. H., L. Madonia, et al. (1999). "Motherhood improves learning and memory." *Nature* 402 (6758): 137–38.

Kinsley, C. H., R. Trainer, et al. (2006). "Motherhood and the hormones of pregnancy modify concentrations of hippocampal neuronal dendritic spines." *Horm Behav* 49 (2): 131–42.

Kirsch, P., C. Esslinger, et al. (2005). "Oxytocin modulates neural circuitry for social cognition and fear in humans." *J Neurosci*, 25

(49): 11489–93.

Kirschbaum, C., B. M. Kudielka, et al. (1999). "Impact of gender, menstrual cycle phase, and oral contraceptives on the activity of the hypothalamus-pituitary-adrenal axis." *Psychosom Med* 61 (2): 154–62.

Klatzkin, R. R., A. L. Morrow, et al. (2006). "Histories of depression, allopregnanolone responses to stress, and premenstrual symptoms in women." *Biol Psychol* 71 (1): 2–11.

Klein, L. C., and E. J. Corwin (2002). "Seeing the unexpected: How sex differences in stress responses may provide a new perspective on the manifestation of psychiatric disorders." *Curr Psychiatry Rep* 4 (6): 441–48.

Knafo, A., A. C. Iervolino, et al. (2005). "Masculine girls and feminine boys: genetic and environmental contributions to atypical gender development in early childhood." *J Pers Soc Psychol* 88 (2): 400–12.

Knaus, T. A., A. M. Bollich, et al. (2004). "Sex-linked differences in the anatomy of the perisylvian language cortex: A volumetric MRI study of gray matter volumes." *Neuropsychology* 18 (4): 738–47.

Knaus, T. A., A. M. Bollich, et al. (2006). "Variability in perisylvian brain anatomy in healthy adults." *Brain Lang.* In press.

Knickmeyer, R., S. Baron-Cohen, et al. (2005). "Foetal testosterone, social relationships, and restricted interests in children." *J Child Psychol Psychiatry* 46 (2): 198–210.

Knickmeyer, R. C., S. Wheelwright, et al. (2005). "Gender-typed play and amniotic testosterone." *Dev Psychol* 41 (3): 517–28.

Knight, G., I. Gunthrie, et al. (2002). "Emotional arousal and gender differences in aggression: A meta-analysis." *Aggressive Behavior* 28:366–93.

Koch, P. (2005). "Feeling Frumpy": The relationships between body image and sexual response changes in midlife women." *J Sex Res* 42 (3) 212–19.

Kochanska, G., K. DeVet, et al. (1994). "Maternal reports of conscience development and temperament in young children." *Child Dev* 65 (3): 852–68.

Kochunov, P., J. F. Mangin, et al. (2005). "Age-related morphology trends of cortical sulci." *Hum Brain Mapp* 26 (3): 210–20.

Komesaroff, P. A., M. D. Esler, et al. (1999). "Estrogen

supplementation attenuates glucocorticoid and catecholamine responses to mental stress in perimenopausal women." *J Clin Endocrinol Metab* 84 (2): 606–10.

Korol, D. L. (2004). "Role of estrogen in balancing contributions from multiple memory systems." *Neurobiol Learn Mem* 82 (3): 309–23.

Korol, D. L., E. L. Malin, et al. (2004). "Shifts in preferred learning strategy across the estrous cycle in female rats." *Horm Behav* 45 (5): 330–38.

Kosfeld, M., M. Heinrichs, et al. (2005). "Oxytocin increases trust in humans." *Nature* 435 (7042): 673–76.

Kravitz, H. (2005). "Relationship of day-to-day reproductive levels to sleep in midlife women." *Arch Intern Med* 165:2370–76.

Kring, A. M., (2000). "Gender and anger." In *Gender and Emotion: Social Psychological Perspectives: Studies in Emotion and Social Interaction*, ed. A. H. Fischer, 2nd series (211–31). New York: Cambridge University Press.

Kring, A. M. (1998). "Sex differences in emotion: Expression, experience, and physiology" *J Pers Soc Psychol* 74 (3): 686–703.

Krpan, K. M., R. Coombs, et al. (2005). "Experiential and hormonal correlates of maternal behavior in teen and adult mothers." *Horm Behav* 47 (1): 112–22.

Krueger, R. B., and M. S. Kaplan (2002). "Treatment resources for the paraphilic and hypersexual disorders." *J Psychiatr Pract* 8 (1): 59–60.

Kruijver, F. P., A. Fernandez-Guasti, et al. (2001). "Sex differences in androgen receptors of the human mamillary bodies are related to endocrine status rather than to sexual orientation or transsexuality." *J Clin Endocrinol Metab* 86 (2): 818–27.

Kudielka, B. M., A. Buske-Kirschbaum, et al. (2004). "HPA axis responses to laboratory psychosocial stress in healthy elderly adults, younger adults, and children: Impact of age and gender." *Psychoneuroendocrinology* 29 (1): 83–98.

Kudielka, B. M., and C. Kirschbaum (2005). "Sex differences in HPA axis responses to stress: A review." *Biol Psychol* 69 (1): 113–32.

Kudielka, B. M., A. K. SchmidtReinwald, et al. (1999). "Psychological and endocrine responses to psychosocial stress and dexamethasone/corticotropin-releasing hormone in healthy postmenopausal women and young controls: The impact of age and a two-week estradiol treatment." *Neuroendocrinology* 70 (6): 422–30.

Kuhlmann, S., C. Kirschbaum, et al. (2005). "Effects of oral cortisol treatment in healthy young women on memory retrieval of negative and neutral words." *Neurobiol Learn Mem* 83 (2): 158–62.

Kuhlmann, S., and O. T. Wolf (2005). "Cortisol and memory retrieval in women: influence of menstrual cycle and oral contraceptives." *Psychopharmacology (Berl)* 183 (1): 65–71.

Kurosaki, M., N. Shirao, et al. (2006). "Distorted images of one's own body activates the prefrontal cortex and limbic/paralimbic system in young women: A functional magnetic resonance imaging study." *Biol Psychiatry*. In press.

Kurshan, N., and C. Neill Epperson (2006). "Oral contraceptives and mood in women with and without premenstrual dysphoria: A theoretical model." *Arch Women Ment Health* 9 (1): 1–14.

Labouvie-Vief, G., M. A. Lumley, et al. (2003). "Age and gender differences in cardiac reactivity and subjective emotion responses to emotional autobiographical memories." *Emotion* 3 (2): 115–26.

Ladd, C. O., D. J. Newport, et al. (2005). "Venlafaxine in the treatment of depressive and vasomotor symptoms in women with peri-menopausal depression." *Depress Anxiety* 22 (2): 94–97.

Lakoff, R. (1976). *Language and Women's Place*. New York: Harper & Row.

Lambert, K. G., A. E. Berry, et al. (2005). "Pup exposure differentially enhances foraging ability in primiparous and nulliparous rats." *Physiol Behav* 84 (5): 799–806.

Laumann, E. O., A. Nicolosi, et al. (2005). "Sexual problems among women and men aged 40–80: Prevalence and correlates identified in the Global Study of Sexual Attitudes and Behaviors." *Int J Impot Res* 17 (1): 39–57.

Laumann, E. O., A. Paik, et al. (1999). "Sexual dysfunction in the United States: Prevalence and predictors." *JAMA* 281 (6): 537–44.

Lavelli, M., and A. Fogel (2002). "Developmental changes in mother-infant face-to-face communication: Birth to 3 months." *Dev Psychol* 38 (2): 288–305.

Lawal, A., M. Kern, et al. (2005). "Cingulate cortex: A closer look at its gut-related functional topography." *Am J Physiol Gastrointest Liver Physiol* 289 (4): G722–30.

Lawrence, P. (2006). "Men, women and ghosts in science." *PLoS Biology* 4 (1): 19.

Lawrence, P. A. (2003). "The politics of publication." *Nature* 422 (6929): 259–61.

Leaper, C., and T. E. Smith (2004). "A meta-analytic review of gender variations in children's language use: Talkativeness, affiliative speech, and assertive speech." *Dev Psychol* 40 (6): 993–1027.

Leckman, J. F., R. Feldman, et al. (2004). "Primary parental preoccupation: Circuits, genes, and the crucial role of the environment." *J Neural Transm* 111 (7): 753–71.

Leckman, J. F., and L. C. Mayes (1999). "Preoccupations and behaviors associated with romantic and parental love: Perspectives on the origin of obsessive-compulsive disorder." *Child Adolesc Psychiatr Clin N Am* 8 (3): 635–65.

Lederman, S. A. (2004). "Influence of lactation on body weight regulation." *Nutr Rev* 62 (7, Pt. 2): S112–19.

Lederman, S. A., V. Rauh, et al. (2004). "The effects of the World Trade Center event on birth outcomes among term deliveries at three lower Manhattan hospitals." *Environ Health Perspect* 112 (17): 1772–78.

Lee, M., U. F. Bailer, et al. (2005). "Relationship of a 5-HT transporter functional polymorphism to 5-HT1A receptor binding in healthy women." *Mol Psychiatry* 10 (8): 715–16.

Lee, T. M., H. L. Liu, et al. (2002). "Gender differences in neural correlates of recognition of happy and sad faces in humans assessed by functional magnetic resonance imaging." *Neurosci Lett* 333 (1): 131–36.

Lee, T. M., H. L. Liu, et al. (2005). "Neural activities associated with emotion recognition observed in men and women." *Mol Psychiatry* 10 (5): 450–55.

Leeb, R. T. R., and F. Gillian (2004). "Here's looking at you, kid! A longitudinal study of perceived gender differences in mutual gaze behavior in young infants." *Sex Roles* 50 (1–2): 1–5.

Legato, M. J. (2005). "Men, women, and brains: What's hardwired, what's learned, and what's controversial." *Gend Med* 2 (2): 59–61.

Leibenluft, E., M. I. Gobbini, et al. (2004). "Mothers' neural activation in response to pictures of their children and other children." *Biol Psychiatry* 56 (4): 225–32.

Leppänen, J. M. H. (2001). "Emotion recognition and social adjustment in school-aged girls and boys." *Scand J Psychol* 42 (5): 429–35.

Leresche, L., L. A. Mancl, et al. (2005). "Relationship of pain and symptoms to pubertal development in adolescents." *Pain* 118 (1–2): 201–9.

LeVay, S. (1991). "A difference in hypothalamic structure between heterosexual and homosexual men." *Science* 253 (5023): 1034–37.

Levenson, R. W. (2003). "Blood, sweat, and fears: The autonomic architecture of emotion." *Ann NY Acad Sci* 1000:348–66.

Levesque, J., F. Eugene, et al. (2003). "Neural circuitry underlying voluntary suppression of sadness." *Biol Psychiatry* 53 (6): 502–10.

Levesque, J., Y. Joanette, et al. (2003). "Neural correlates of sad feelings in healthy girls." *Neuroscience* 121 (3): 545–51.

Lewis, D. A., D. Cruz, et al. (2004). "Postnatal development of prefrontal inhibitory circuits and the pathophysiology of cognitive dysfunction in schizophrenia." *Ann NY Acad Sci* 1021:64–76.

Lewis, M. (1997). "Social behavior and language acquisition." In *Interactional conversation and the development of language*, ed. B. Haslett, New York: Wiley. 313–30.

Li, C. S., T. R. Kosten, et al. (2005). "Sex differences in brain activation during stress imagery in abstinent cocaine users: A functional magnetic resonance imaging study." *Biol Psychiatry* 57 (5): 487–94.

Li, H., S. Pin, et al. (2005). "Sex differences in cell death." *Ann Neurol* 58 (2): 317–21.

Li, L., E. B. Keverne, et al. (1999). "Regulation of maternal behavior and offspring growth by paternally expressed Peg3." *Science* 284 (5412): 330–33.

Li, M., and A. S. Fleming (2003). "The nucleus accumbens shell is critical for normal expression of pup-retrieval in postpartum female rats." *Behav Brain Res* 145 (1–2): 99–111.

Li, R., and Y. Shen (2005). "Estrogen and brain: Synthesis, function and diseases." *Front Biosci* 10: 257–67.

Li, Z. J., H. Matsuda, et al. (2004). "Gender difference in brain perfusion 99mTc-ECD SPECT in aged healthy volunteers after correction for partial volume effects." *Nucl Med Commun* 25 (10): 999–1005.

Light, K. C., K. M. Grewen, et al. (2004). "Deficits in plasma oxytocin responses and increased negative affect, stress, and blood pressure in mothers with cocaine exposure during pregnancy." *Addict Behav* 29 (8): 1541–64.

Light, K. C., K. M. Grewen, et al. (2005). "More frequent partner hugs and higher oxytocin levels are linked to lower blood pressure and heart rate in premenopausal women." *Biol Psychol* 69 (1): 5–21.

Light, K. C., K. M. Grewen, et al. (2005). "Oxytocinergic activity is linked to lower blood pressure and vascular resistance during stress in postmenopausal women on estrogen replacement." *Horm Behav* 47 (5): 540–48.

Light, K. C., T. E. Smith, et al. (2000). "Oxytocin responsivity in mothers of infants: A preliminary study of relationships with blood pressure during laboratory stress and normal ambulatory activity." *Health Psychol* 19 (6): 560–67.

Lim, M. M., I. F. Bielsky, et al. (2005). "Neuropeptides and the social brain: Potential rodent models of autism." *Int J Dev Neurosci* 23 (2–3): 235–43.

Lim, M. M., E. A. Hammock, et al. (2004). "The role of vasopressin in the genetic and neural regulation of monogamy." *J Neuroendocrinol* 16 (4): 325–32.

Lim, M. M., A. Z. Murphy, et al. (2004). "Ventral striatopallidal oxytocin and vasopressin V1a receptors in the monogamous prairie vole (Microtus ochrogaster)." *J Comp Neurol* 468 (4): 555–70.

Lim, M. M., H. P. Nair, et al. (2005). "Species and sex differences in brain distribution of corticotropin-releasing factor receptor subtypes 1 and 2 in monogamous and promiscuous vole species." *J Comp Neurol* 487 (1): 75–92.

Lim, M. M., Z. Wang, et al. (2004). "Enhanced partner preference in a promiscuous species by manipulating the expression of a single gene." *Nature* 429 (6993): 754–57.

Lim, M. M., and L. J. Young (2004). "Vasopress-independent neural circuits underlying pair bond formation in the monogamous prairie vole." *Neuroscience* 125 (1): 35–45.

Lobo, R. (2000). *Menopause.* San Diego: Academic Press.

Lobo, R. A. (2005). "Appropriate use of hormones should alleviate concerns of cardiovascular and breast cancer risk." *Maturitas* 51 (1): 98–109.

Logsdon, M. C., K. Wisner, et al. (2006). "Raising the awareness of primary care providers about postpartum depression." *Issues Ment Health Nurs* 27 (1): 59–73.

Lonstein, J. S. (2005). "Reduced anxiety in postpartum rats requires recent physical interactions with pups, but is independent of suckling and peripheral sources of hormones." *Horm Behav* 47 (3): 241–55.

Lovell-Badge, R. (2005). "Aggressive behaviour: Contributions from genes on the Y chromosome." *Novartis Found Symp* 268:20–33; discussion 33–41, 96–99.

Lovic, V., and A. S. Fleming (2004). "Artificially-reared female rats show reduced prepulse inhibition and deficits in the attentional set shifting task—reversal of effects with material-like licking stimulation." *Behav Brain Res* 148 (1–2): 209–19.

Lu, N. Z., and C. L. Bethea (2002). "Ovarian steroid regulation of 5-HT1A receptor binding and G protein activation in female monkeys." *Neuropsychopharmacology* 27 (1): 12–24.

Luisi, A. F., and J. E. Pawasauskas (2003). "Treatment of premenstrual dysphoric disorder with selective serotonin reuptake inhibitors." *Pharmacotherapy* 23 (9): 1131–40.

Luna, B. (2004). "Algebra and the adolescent brain." *Trends Cogn Sci* 8 (10): 437–39.

Luna, B., K. E. Garver, et al. (2004). "Maturation of cognitive processes from late childhood to adulthood." *Child Dev* 75 (5): 1357–72.

Lunde, I., G. K. Larson, et al. (1991). "Sexual desire, orgasm, and sexual fantasies: A study of 625 Danish women born in 1910, 1936 and 1958." *J Sex Educ Ther*, 17:62–70.

Lundstrom, J. N., M. Goncalves, et al. (2003). "Psychological effects of subthreshold exposure to the putative human pheromone 4,16-androstadien-3-one." *Horm Behav* 44 (5): 395–401.

Lynam, D. (2004). "Personality pathways to impulsive behavior and their relations to deviance: Results from three samples." *Journal of Quantitative Criminology* 20:319–41.

McCarthy, M. M., C. H. McDonald, et al. (1996). "An anxiolytic action of oxytocin is enhanced by estrogen in the mouse." *Physiol Behav* 60 (5): 1209–15.

McClintock, M. (2002). "Pheromones, odors and vsana: The neuroendocrinology of social chemosignals in humans and animals." In *Hormones, Brain and Behavior*, ed. D. W. Pfaff, vol. 1, 797–870.

McClintock, M. K. (1998). "On the nature of mammalian and human pheromones." *Ann NY Acad Sci* 855:390–92.

McClintock, M. K., S. Bullivant, et al. (2005). "Human body scents: Conscious perceptions and biological effects." *Chem Senses* 30 (Suppl. 1): i135–i137.

McClure, E. B. (2000). "A meta-analytic review of sex differences in facial expression processing and their development in infants, children, and adolescents." *Psychol Bull* 126 (3): 424–53.

McClure, E. B., C. S. Monk, et al. (2004). "A developmental examination of gender differences in brain engagement during evaluation of threat." *Biol Psychiatry* 55 (11): 1047–55.

Maccoby, E. E. (1959). "Role-taking in childhood and its consequences for social learning." *Child Dev* 30 (2): 239–52.

Maccoby, E. E. (1998). *The Two Sexes: Growing Up Apart, Coming Together*. Cambridge, MA: Harvard University Press.

Maccoby, E. E. (2005). Personal communication.

Maccoby, E. E., and C. N. Jacklin (1973). "Stress, activity, and proximity seeking: Sex differences in the year-old child." *Child Dev* 44 (1): 34–42.

Maccoby, E. E., and C. N. Jacklin (1980). "Sex differences in aggression: A rejoinder and reprise." *Child Dev* 51 (4): 964–80.

Maccoby, E. E., and C. N. Jacklin (1987). "Gender segregation in childhood." *Adv Child Dev Behav* 20:239–87.

McCormick, C. M., and E. Mahoney (1999). "Persistent effects of prenatal, neonatal, or adult treatment with flutamide on the hypothalamic-pituitary-adrenal stress response of adult male rats." *Horm Behav* 35 (1): 90–101.

McEwen, B. S. (2001). "Invited review: Estrogen's effects on the brain: Multiple sites and molecular mechanisms." *J Appl Physiol* 91 (6): 2785–801.

McEwen, B. S., and J. P. Olie (2005). "Neurobiology of mood, anxiety, and emotions as revealed by studies of a unique antidepressant: Tianeptine." *Mol Psychiatry* 10 (6): 525–37.

McFadden, D., and E. G. Pasanen (1998). "Comparison of the auditory systems of heterosexuals and homosexuals: Click-evoked otoacoustic emissions." *Proc Natl Acad Sci USA* 95 (5): 2709–13.

McFadden, D., and E. G. Pasanen (1999). "Spontaneous otoacoustic emissions in heterosexuals, homosexuals, and bisexuals." *J Acoust Soc Am* 105 (4): 2403–13.

McGinnis, M. Y. (2004). "Anabolic androgenic steroids and aggression: Studies using animal models." *Ann NY Acad Sci* 1036:399–415.

McManis, M. H., M. M. Bradley, et al. (2001). "Emotional reactions in children: Verbal, physiological, and behavioral responses to affective pictures." *Psychophysiology* 38 (2): 222–31.

Maciejewski, P. K., H. G. Prigerson, et al. (2001). "Sex differences in event-related risk for major depression." *Psychol Med* 31 (4): 593–604.

Mackey, R., (2001). "Psychological intimacy in the lasting relationships of heterosexual and same-gender couples." *Sex Roles* 43 (3–4): 201.

Mackie, D. M., T. Devos, et al. (2000). "Intergroup emotions: Explaining offensive action tendencies in an intergroup context." *J Pers Soc Psychol* 79 (4): 602–16.

Madden, T. E., L. F. Barrett, et al. (2000). "Sex differences in anxiety and depression: Empirical evidence and methodological questions." In *Gender and Emotion: Social Psychological Perspectives: Studies in Emotion and Social Interaction*, ed. A. H. Fischer, 2nd series, 277–98. New York: Cambridge University Press.

Maestripieri, D. (2005). "Early experience affects the intergenerational transmission of infant abuse in rhesus monkeys." *Proc Natl Acad Sci USA* 102 (27): 9726–29.

Maestripieri, D. (2005). "Effects of early experience on female behavioural and reproductive development in rhesus macaques." *Proc Biol Sci* 272 (1569): 1243–48.

Maestripieri, D., S. G. Lindell, et al. (2005). "Neurobiological characteristics of rhesus macaque abusive mothers and their relation to social and maternal behavior." *Neurosci Biobehav Rev* 29 (1): 51–57.

Magalhaes, P. V., and R. T. Pinheiro (2006). "Pharmacological treatment of postpartum depression." *Acta Psychiatr Scand* 113 (1): 75–76.

Maki, P. M., A. B. Zonderman, et al. (2001). "Enhanced verbal memory in nondemented elderly women receiving hormone-replacement therapy." *Am J Psychiatry* 158 (2): 227–33.

Malatesta, C. Z., and J. M. Haviland (1982). "Learning display rules: The socialization of emotion expression in infancy." *Child Dev* 53 (4): 991–1003.

Mandal, M. K. (1985). "Perception of facial affect and physical proximity." *Percept Mot Skills* 60 (3): 782.

Mani, S. (2002). "Mechanisms of progesterone receptor action in the brain." In *Hormones, Brain and Behavior*, ed. D. W. Pfaff, vol. 3, 643–82. San Diego, Academic Press.

Mann, P. E., and J. A. Babb (2005). "Neural steroid hormone receptor gene expression in pregnant rats." *Brain Res Mol Brain Res* 142 (1): 39–46.

Manning, J. T., A. Stewart, et al. (2004). "Sex and ethnic differences in 2nd to 4th digit ratio of children." *Early Hum Dev* 80 (2): 161–68.

Marshall, E. (2005). "From dearth to deluge." *Science* 308 (5728): 1570.

Martel, F. L., C. M. Nevison, et al. (1993). "Opioid receptor blockade reduces maternal affect and social grooming in rhesus monkeys." *Psychoneuroendocrinology* 18 (4): 307–21.

Martin-Loeches, M., R. M. Orti, et al. (2003). "A comparative analysis of the modification of sexual desire of users of oral hormonal contraceptives and intrauterine contraceptive devices." *Eur J Contracept Reprod Health Care* 8 (3): 129–34.

Masoni, S., A. Maio, et al. (1994). "The couvade syndrome." *J Psychosom Obstet Gynaecol* 15 (3): 125–31.

Mass, J. (1998). Sleep: *The Revolutionary Program that Prepares Your Mind for Peak Performance.* New York: HarperCollins.

Mathews, G. A., B. A. Fane, et al. (2004). "Androgenic influences on neural asymmetry: Handedness and language lateralization in individuals with congenital adrenal hyperplasia." *Psychoneuroendocrinology* 29 (6): 810–22.

Matthews, T. J., P. Abdelbaky, et al. (2005). "Social and sexual motivation in the mouse." *Behav Neurosci* 119 (6): 1628–39.

Matthiesen, A. S., A. B. Ransjo-Arvidson, et al. (2001). "Postpartum maternal oxytocin release by newborns: Effects of infant hand massage and sucking." *Birth* 28 (1): 13–19.

Mazure, C. M., and P. K. Maciejewski (2003). "A model of risk for major depression: Effects of life stress and cognitive style vary by age." *Depress Anxiety* 17 (1): 26–33.

Meaney, M. (2001). "From a culture of blame to a culture of safety—the role of institutional ethics committees." *Bioethics Forum* 17 (2): 32–42.

Meaney, M. J. (2001). "Maternal care, gene expression, and the transmission of individual differences in stress reactivity across generations." *Annu Rev Neurosci* 24:1161–92.

Meaney, M. J., and M. Szyf (2005). "Maternal care as a model for experience-dependent chromatin plasticity?" *Trends Neurosci* 28 (9): 456–63.

Mellon, S., L. Brizendine and S. Conrad, (2004). "Neurosteroids, PMS and depression." *Behavioral Pharmacology* 15:22–28.

Mellon, S., S. Conrad, et al. (2006). "Allopregnanolone synthesis vs cycle vs normal vs PMDD." In preparation.

Mendelsohn, M. E., and R. H. Karas (2005). "Molecular and cellular basis of cardiovascular gender differences." *Science* 308 (5728): 1583–87.

Mendoza, E., and G. Carballo (1999). "Vocal tremor and psychological stress." *J Voice* 13 (1): 105–12.

Mendoza, S. P. (1999). "Attachment relationships in New World primates." In C. S. Carter, I. I. Lederhendler, and B. Kirkpatrick, eds., *The Integrative Neurobiology of Affiliation*, 93–100. Cambridge, MA: MIT Press.

Miller, G. E., N. Rohleder, et al. (2006). "Clinical depression and regulation of the inflammatory response during acute stress." *Psychosom Med* In press.

Miller, K. J., J. C. Conney, et al. (2002). "Mood symptoms and cognitive performance in women estrogen users and nonusers and men." *J Am Geriatr Soc* 50 (11): 1826–30.

Miller, S. M., and J. S. Lonstein (2005). "Dopamine d1 and d2 receptor antagonism in the preoptic area produces different effects on maternal behavior in lactating rats." *Behav Neurosci* 119 (4): 1072–83.

Mitchell, J. P., M. R. Banaji, et al. (2005). "The link between social cognition and self-referential thought in the medial prefrontal cortex." *J Cogn Neurosci* 17 (8): 1306–15.

Moffitt, T. (2001). *Sex Differences in Antisocial Behavior.* Cambridge: Cambridge University Press.

Mogi, K., T. Funabashi, et al. (2005). "Sex difference in the response of melaninconcentrating hormone neurons in the lateral hypothalamic area to glucose, as revealed by the expression of phosphorylated cyclic adenosine 3', 5'monophosphate response element-binding protein." *Endocrinology* 146 (8): 3325–33.

Monks, D. A., J. S. Lonstein, et al. (2003). "Got milk? Oxytocin triggers hippocampal plasticity." *Nat Neurosci* 6 (4): 327–28.

Monnet, F. P., and T. Maurice (2006). "The sigma(1) protein as a target for the nongenomic effects of neuro(active) steroids: Molecular, physiological, and behavioral aspects." *J Pharmacol Sci* 100 (2): 93–118.

Morgan, H. D., A. S. Fleming, et al. (1992). "Somatosensory control of the onset and retention of maternal responsiveness in primiparous Sprague-Dawley rats." *Physiol Behav* 51 (3): 549–55.

Morgan, M. A., J. Schulkin, et al. (2004). "Estrogens and non-reproductive behaviors related to activity and fear." *Neurosci Biobehav Rev* 28 (1): 55–63.

Morgan, M. L., I. A. Cook, et al. (2005). "Estrogen augmentation of antidepressants in perimenopausal depression: A pilot study." *J Clin Psychiatry* 66 (6): 774–80.

Morley-Fletcher, S., M. Puopolo, et al. (2004). "Prenatal stress affects 3, 4-methylenedioxymethamphetamine pharmacokinetics and druginduced motor alterations in adolescent female rats." *Eur J Pharmacol* 489 (1–2): 89–92.

Morley-Fletcher, S., M. Rea, et al. (2003). "Environmental enrichment during adolescence reverses the effects of prenatal stress on play behaviour and HPA axis reactivity in rats." *Eur J Neurosci* 18 (12): 3367–74.

Morse, C. A., and K. Rice (2005). "Memory after menopause: Preliminary considerations of hormone influence on cognitive functioning." *Arch Women Ment Health* 8 (3): 155–62.

Motzer, S. A., and V. Hertig (2004). "Stress, stress response, and health." *Nurs Clin North Am* 39 (1): 1–17.

Mowlavi, A., D. Cooney, et al. (2005). "Increased cutaneous nerve fibers in female specimens." *Plast Reconstr Surg* 116 (5): 1407–10.

Muller, M., D. E. Grobbee, et al. (2005). "Endogenous sex hormones and metabolic syndrome in aging men." *J Clin Endocrinol Metab* 90 (5): 2618–23.

Muller, M., M. E. Keck, et al. (2002). "Genetics of endocrine-behavior interactions." In *Hormones, Brain and Behavior*, ed. D. W. Pfaff, vol. 5, 263–302, San Diego: Academic Press.

Murabito, J. M., Q. Yang, et al. (2005). "Heritability of age at natural menopause in the Framingham Heart Study." *J Clin Endocrinol Metab* 90 (6): 3427–30.

Murphy, C. T., S. A. McCarroll, et al. (2003). "Genes that act downstream of DAF-16 to influence the lifespan of *Caenorhabditis elegans*." *Nature* 424 (6946): 277–83.

Muscarella, F., V. A. Elias, et al. (2004). "Brain differentiation and preferred partner characteristics in heterosexual and homosexual men and women." *Neuro Endocrinol Lett* 25 (4): 297–301.

Must, A., E. N. Naumova, et al. (2005). "Childhood overweight and maturational timing in the development of adult overweight and fatness: The Newton Girls Study and its follow-up." *Pediatrics* 116 (3): 620–27.

Mustanski, B. S., M. G. Dupree, et al. (2005). "A genomewide scan of male sexual orientation." *Hum Genet* 116 (4): 272–78.

Naftolin, F. (2005). "Prevention during the menopause is critical for good health: Skin studies support protracted hormone therapy." *Fertil Steril* 84 (2): 293–94; discussion 295.

Nagy, E. (2001). "Different emergence of fear expression in infant boys and girls." *Infant Behavior and Development* 24:189–94.

Naliboff, B. D., S. Berman, et al. (2003). "Sex-related differences in IBS patients: Central processing of visceral stimuli."

Gastroenterology 124 (7): 1738–47.

Nawata, H., T. Yanase, et al. (2004). "Adrenopause." *Horm Res* 62 (Suppl. 3): 110–14.

Neff, B. D. (2003). "Decisions about parental care in response to perceived paternity." *Nature* 422 (6933): 716–19.

Neighbors, K. A., B. Gillespie, et al. (2003). "Weaning practices among breastfeeding women who weaned prior to six months postpartum." *J Hum Lact* 19 (4): 374–80; quiz 381–5, 448.

Nelson, E. E., E. Leibenluft, et al. (2005). "The social re-orientation of adolescence: A neuroscience perspective on the process and its relation to psychopathology." *Psychol Med* 35 (2): 163–74.

Netherton, C., I. Goodyer, et al. (2004). "Salivary cortisol and dehydroepiandrosterone in relation to puberty and gender." *Psychoneuroendocrinology* 29 (2): 125–40.

Niederle, M. (2005). "Why do women shy away from competition? Do men compete too much?" *NBER*, working paper, July 2005.

Nishida, Y., M. Yoshioka, et al. (2005). "Sexually dimorphic gene expression in the hypothalamus, pituitary gland, and cortex." *Genomics* 85 (6): 679–87.

Nitschke, J. B., E. E. Nelson, et al. (2004). "Orbitofrontal cortex tracks positive mood in mothers viewing pictures of their newborn infants." *Neuroimage* 21 (2): 583–92.

Oatridge, A., A. Holdcroft, et al. (2002). "Change in brain size during and after pregnancy: Study in healthy women and women with preeclampsia." *AJNR Am J Neuroradiol* 23 (1): 19–26.

Oberman, L. M. (2005). Personal communication: "There may be a difference in male and female mirror neuron functioning."

Oberman, L. M., E. M. Hubbard, et al. (2005). "EEG evidence for mirror neuron dysfunction in autism spectrum disorders." *Brain Res Cogn Brain Res* 24 (2): 190–98.

Ochsner, K. N., R. D. Ray, et al. (2004). "For better or for worse: Neural systems supporting the cognitive down and up-regulation of negative emotion." *Neuroimage* 23 (2): 483–99.

O'Connell, H. E., K. V. Sanjeevan, et al. (2005). "Anatomy of the clitoris." *J Urol* 174 (4, Pt. 1): 1189–95.

O'Connor, D. B., J. Archer, et al. (2004). "Effects of testosterone on mood, aggression, and sexual behavior in young men: A double-blind, placebo-controlled, crossover study." *J Clin Endocrinol Metab* 89 (6): 2837–45.

O'Day, D. H., M. Lydan, et al. (2001). "Decreases in calmodulin binding proteins and calmodulin dependent protein phosphorylation in the medial preoptic area at the onset of maternal behavior in the rat." *J Neurosci Res* 64 (6): 599–605.

O'Day, D. H., L. A. Payne, et al. (2001). "Loss of calcineurin from the medial preoptic area of primiparous rats." *Biochem Biophys Res Commun* 281 (4): 1037–40.

O'Hara, M. W., J. A. Schlechte, et al. (1991). "Controlled prospective study of postpartum mood disorders: Psychological, environmental, and hormonal variables." *J Abnorm Psychol* 100 (1): 63–73.

O'Hara, M. W., J. A. Schlechte, et al. (1991). "Prospective study of postpartum blues: Biologic and psychosocial factors." *Arch Gen Psychiatry* 48 (9): 801–6.

Ohnishi, T., Y. Moriguchi, et al. (2004). "The neural network for the mirror system and mentalizing in normally developed children: An fMRI study." *Neuroreport* 15 (9): 1483–87.

Ojeda, S. (2002). "Neuroendocrine regulation of puberty." In *Hormones, Brain and Behavior*, ed. D. W. Pfaff, vol. 4, 589–660. San Diego, Academic Press.

Olweus, D., A. Mattsson, et al. (1988). "Circulating testosterone levels and aggression in adolescent males: A causal analysis." *Psychosom Med* 50 (3): 261–72.

OpenSpeechRecognizer (2005). "Male and female spectral tones of voice." See www.nuance.com.

Orzhekhovskaia, N. S. (2005). "[Sex dimorphism of neuronglia correlations in the frontal areas of the human brain]." *Morfologiia* 127 (1): 7–9.

Otte, C., S. Hart, et al. (2005). "A meta-analysis of cortisol response to challenge in human aging: Importance of gender." *Psychoneuroendocrinology* 30 (1): 80–91.

Overman, W. H., J. Bachevalier, et al. (1996). "Cognitive gender differences in very young children parallel biologically based cognitive gender differences in monkeys." *Behav Neurosci* 110 (4): 673–84.

Palermo, R. C. (2004). "Photographs of facial expression: Accuracy, response times, and ratings of intensity." *Behavior Research Methods, Instruments & Computers.* Special Web-based archive of norms, stimuli, and data, Pt. 2, 36 (4): 634–38.

Panzer, C., S. Wise, et al. (2006). "Impact of oral contraceptives on sex hormone-binding globulin and androgen levels: A retrospective study in women with sexual dysfunction." *J Sex Med* 3 (1): 104–13.

Papalexi, E., K. Antoniou, et al. (2005). "Estrogens influence behavioral responses in a kainic acid model of neurotoxicity." *Horm Behav* 48 (3): 291–302.

Paris, R., and R. Helson (2002). "Early mothering experience and personality change." *J Fam Psychol* 16 (2): 172–85.

Parry, B. (2002). "Premenstrual dysphoric disorder PMDD." In *Hormones, Brain and Behavior,* ed. D. W. Pfaff, vol. 5, 531–52. San Diego, Academic Press.

Parsey, R. V., M. A. Oquendo, et al. (2002). "Effects of sex, age, and aggressive traits in man on brain serotonin 5-HT1A receptor binding potential measured by PET using [C11]WAY-100635." *Brain Res* 954 (2): 173–82.

Pasterski, V. L., M. E. Geffner, et al. (2005). "Prenatal hormones and postnatal socialization by parents as determinants of male-typical toy play in girls with congenital adrenal hyperplasia." *Child Dev* 76 (1): 264–78.

Pattatucci, A. M., and D. H. Hamer (1995). "Development and familiality of sexual orientation in females." *Behav Genet* 25 (5): 407–20.

Paus, T., A. Zijdenbos, et al. (1999). "Structural maturation of neural pathways in children and adolescents: In vivo study." *Science* 283 (5409): 1908–11.

Pawluski, J. L., and L. A. Galea (2006). "Hippocampal morphology is differentially affected by reproductive experience in the mother." *J Neurobiol* 66 (1): 71–81.

Pawluski, J. L., S. K. Walker, et al. (2006). "Reproductive experience differentially affects spatial reference and working memory performance in the mother." *Horm Behav* 49 (2): 143–49.

Pazol, K., K. V. Northcutt, et al. (2005). "Medroxyprogesterone acetate acutely facilitates and sequentially inhibits sexual behavior in female rats." *Horm Behav*

Pease, A. (1997). *Talk Language*. Sydney: Camel Publishing.

Pedersen, C. A., and M. L. Boccia (2003). "Oxytocin antagonism alters rat dams' oral grooming and upright posturing over pups." *Physiol Behav* 80 (2–3): 233–41.

Pennebaker, J. W., C. J. Groom, et al. (2004). "Testosterone as a social inhibitor: Two case studies of the effect of testosterone treatment on language." *J Abnorm Psychol* 113 (1): 172–75.

Perez-Martin, M., V. Salazar, et al. (2005). "Estradiol and soy extract increase the production of new cells in the dentate gyrus of old rats." *Exp Gerontol* 40 (5): 450–53.

Pezawas, L., A. MeyerLindenberg, et al. (2005). "5-HTTLPR polymorphism impacts human cingulate-amygdala interactions: A genetic susceptibility mechanism for depression." *Nat Neurosci* 8 (6): 828–34.

Phelps, E. A. (2004). "Human emotion and memory: Interactions of the amygdala and hippocampal complex." *Curr Opin Neurobiol* 14 (2): 198–202.

Phillips, S. M., and B. B. Sherwin (1992). "Variations in memory function and sex steroid hormones across the menstrual cycle." *Psychoneuroendocrinology* 17 (5): 497–506.

Pierce, M. B., and D. A. Leon (2005). "Age at menarche and adult BMI in the Aberdeen children of the 1950s cohort study." *Am J Clin Nutr* 82 (4): 733–39.

Pillard, R. C., and J. M. Bailey (1995). "A biologic perspective on sexual orientation." *Psychiatr Clin North Am* 18 (1): 71–84.

Pillsworth, E. G., M. G. Haselton, et al. (2004). "Ovulatory shifts in female sexual desire." *J Sex Res* 41 (1): 55–65.

Pinaud, R., A. F. Fortes, et al. (2006). "Calbindin-positive neurons reveal a sexual dimorphism within the songbird analogue of the mammalian auditory cortex." *J Neurobiol* 66 (2): 182–95.

Pinna, G., E. Costa, et al. (2005). "Changes in brain testosterone and allopregnanolone biosynthesis elicit aggressive behavior." *Proc Natl Acad Sci USA* 102 (6): 2135–40.

Pittman, Q. J., and S. J. Spencer (2005). "Neurohypophysial peptides: Gatekeepers in the amygdala." *Trends Endocrinol Metab* 16 (8): 343–44.

Plante, E., V. J. Schmithorst, et al. (2006). "Sex differences in the activation of language cortex during childhood." *Neuropsychologia.* In press.

Podewils, L. J., E. Guallar, et al. (2005). "Physical activity, APOE genotype, and dementia risk: Findings from the Cardiovascular Health Cognition Study." *Am J Epidemiol* 161 (7): 639–51.

Prkachin, K. M., M. Heather; and S. R. Mercer (2004). "Effects of exposure on perception of pain expression." *Pain* 111 (1–2): 8–12.

Protopopescu, X., H. Pan, et al. (2005). "Orbitofrontal cortex activity related to emotional processing changes across the menstrual cycle." *Proc Natl Acad Sci USA* 102 (44): 16060–65.

Pruessner, J. C., F. Champagne, et al. (2004). "Dopamine release in response to a psychological stress in humans and its relationship to early life maternal care: A positron emission tomography study using [11C]raclopride." *J Neurosci* 24 (11): 2825–31.

Pujol, J., A. Lopez, et al. (2002). "Anatomical variability of the anterior cingulate gyrus and basic dimensions of human personality." *Neuroimage* 15 (4): 847–55.

Putnam, K., G. P. Chrousos, et al. (2005). "Sex-related differences in stimulated hypothalamic-pituitary-adrenal axis during induced gonadal suppression." *J Clin Endocrinol Metab* 90 (7): 4224–31.

Qian, S. Z., Y. Cheng Xu, et al. (2000). "Hormonal deficiency in elderly males." *Int J Androl* 23 (Suppl. 2): 1–3.

Rahman, Q. (2005). "The neurodevelopment of human sexual orientation." *Neurosci Biobehav Rev* 29 (7): 1057–66.

Rahman, Q., S. Abrahams, et al. (2003). "Sexual orientation-related differences in verbal fluency." *Neuropsychology* 17 (2): 240–46.

Rahman, Q., V. Kumari, et al. (2003). "Sexual orientation-related differences in prepulse inhibition of the human startle response." *Behav Neurosci* 117 (5): 1096–102.

Raingruber, B. J. (2001). "Settling into and moving in a climate of care: Styles and patterns of interaction between nurse psychotherapists and clients." *Perspect Psychiatr Care* 37 (1): 15–27.

Rasgon, N. L., C. Magnusson, et al. (2005). "Endogenous and exogenous hormone exposure and risk of cognitive impairment in Swedish twins: a preliminary study." *Psychoneuroendocrinology* 30 (6): 558–67.

REFERENCES

Rasgon, N., S. Shelton, et al. (2005). "Perimenopausal mental disorders: Epidemiology and phenomenology." *CNS Spectr* 10 (6): 471–78.

Ratka, A. (2005). "Menopausal hot flashes and development of cognitive impairment." *Ann NY Acad Sci* 1052:11–26.

Raz, N., F. Gunning-Dixon, et al. (2004). "Aging, sexual dimorphism, and hemispheric asymmetry of the cerebral cortex: Replicability of regional differences in volume." *Neurobiol Aging* 25 (3): 377–96.

Raz, N., K. M. Rodrigue, et al. (2004). "Hormone replacement therapy and age-related brain shrinkage: Regional effects." *Neuroreport* 15 (16): 2531–34.

Reamy, K. J., and S. E. White (1987). "Sexuality in the puerperium: A review." *Arch Sex Behav* 16 (2): 165–86.

Redoute, J., S. Stoleru, et al. (2000). "Brain processing of visual sexual stimuli in human males." *Hum Brain Mapp* 11 (3): 162–77.

Reno, P. L., R. S. Meindl, et al. (2003). "Sexual dimorphism in *Australopithecus afarensis* was similar to that of modern humans." *Proc Natl Acad Sci USA* 100 (16): 9404–9.

Repetti, R. L. (1989). "Effects of daily workload on subsequent behavior during marital interactions: The role of social withdrawal and spouse support." *J Pers Soc Psychol* 57:651–59.

Repetti, R. L. (1997). "The effects of daily job stress on parent behavior with preadolescents." Society for Research in Child Development meeting, Washington, D.C.

Repetti, R. L., S. E. Taylor, et al. (2002). "Risky families: Family social environments and the mental and physical health of offspring." *Psychol Bull* 128 (2): 330–66.

Resnick, S. M., and P. M. Maki (2001). "Effects of hormone replacement therapy on cognitive and brain aging." *Ann NY Acad Sci* 949:203–14.

Rhoden, E. L., and A. Morgentaler (2004). "Risks of testosterone-replacement therapy and recommendations for monitoring." *N Engl J Med* 350 (5): 482–92.

Rhodes, G. (2006). "The evolutionary psychology of facial beauty." *Annu Rev Psychol* 57:199–226.

Rhodes, G., M. Peters, et al. (2005). "Higher-level mechanisms detect facial symmetry." *Proc Biol Sci* 272 (1570): 1379–84.

Richardson, H. N., E. P. Zorrilla, et al. (2006). "Exposure to repetitive versus varied stress during prenatal development generates two distinct anxiogenic and neuroendocrine profiles in adulthood." *Endocrinology.* In press.

Rilling, J. K., J. T. Winslow, et al. (2004). "The neural correlates of mate competition in dominant male rhesus macaques." *Biol Psychiatry* 56 (5): 364–75.

Roalf, D., N. Lowery, et al. (2006). "Behavioral and physiological findings of gender differences in global-local visual processing." *Brain Cogn.* In press.

Roberts, B. W., R. Helson, et al. (2002). "Personality development and growth in women across 30 years: Three perspectives." *J Pers* 70 (1): 79–102.

Robinson, K., and S. E. Maresh (2001). "Mood, marriage, and menopause." *Journal of Counseling Psychology,* 48 (1): 77–84.

Roca, C. A., P. J. Schmidt, and M. Altemus (1998). "Effects of reproductive steroids on the hypothalamicpituitaryadrenal axis response to low dose dexamethasone." Abstract presented at Neuroendocrine Workshop on Stress. New Orleans.

Roca, C. A., P. J. Schmidt, et al. (2003). "Differential menstrual cycle regulation of hypothalamic-pituitary-adrenal axis in women with premenstrual syndrome and controls." *J Clin Endocrinol Metab* 88 (7): 3057–63.

Roenneberg, T., T. Kuehnle, et al. (2004). "A marker for the end of adolescence." *Curr Biol* 14 (24): R1038–39.

Rogan, M. T., K. S. Leon, et al. (2005). "Distinct neural signatures for safety and danger in the amygdala and striatum of the mouse." *Neuron* 46 (2): 309–20.

Rogers, R. D., N. Ramnani, et al. (2004). "Distinct portions of anterior cingulate cortex and medial prefrontal cortex are activated by reward processing in separable phases of decision-making cognition." *Biol Psychiatry* 55 (6): 594–602.

Romeo, R. D., S. J. Lee, et al. (2004). "Differential stress reactivity in intact and ovariectomized prepubertal and adult female rats." *Neuroendocrinology* 80 (6): 387–93.

Romeo, R. D., S. J. Lee, et al. (2004). "Testosterone cannot activate an adult-like stress response in prepubertal male rats."

References

Neuroendocrinology 79 (3): 125–32.

Romeo, R. D., H. N. Richardson, et al. (2002). "Puberty and the maturation of the male brain and sexual behavior: Recasting a behavioral potential." *Neurosci Biobehav Rev* 26 (3): 381–91.

Romeo, R. D., and C. L. Sisk (2001). "Pubertal and seasonal plasticity in the amygdala." *Brain Res* 889 (1–2): 71–77.

Rose, A. B., D. P. Merke, et al. (2004). "Effects of hormones and sex chromosomes on stress-influenced regions of the developing pediatric brain." *Ann NY Acad Sci* 1032:231–33.

Rose, A. J., and K. D. Rudolph (2006). "A review of sex differences in peer relationship processes: potential trade-offs for the emotional and behavioral development of girls and boys." *Psychol Bull* 132 (1): 98–131.

Rosen, W. D., L. B. Adamson, and R. Bakeman. (1992). "An experimental investigation of infant social referencing: Mothers' messages and gender differences." *Dev Psychol* 28 (6): 1172–78.

Rosenblum, L. A., and M. W. Andrews (1994). "Influences of environmental demand on maternal behavior and infant development." *Acta Paediatr Suppl* 397:57–63.

Rosenblum, L. A., J. D. Coplan, et al. (1994). "Adverse early experiences affect noradrenergic and serotonergic functioning in adult primates." *Biol Psychiatry* 35 (4): 221–27.

Rosip, J. C., J. A. Hall (2004). "Knowledge of nonverbal cues, gender, and nonverbal decoding accuracy." *Journal of Nonverbal Behavior, Special Interpersonal Sensitivity*, Pt. 2. 28 (4): 267–86.

Ross, J. L., D. Roeltgen, et al. (1998). "Effects of estrogen on nonverbal processing speed and motor function in girls with Turner's syndrome." *J Clin Endocrinol Metab* 83 (9): 3198–204.

Rossouw, J. E. (2002). "Effect of postmenopausal hormone therapy on cardiovascular risk." *J Hypertens Suppl* 20 (2): S62–65.

Rossouw, J. E. (2002). "Hormones, genetic factors, and gender differences in cardiovascular disease." *Cardiovasc Res* 53 (3): 550–57.

Rossouw, J. E., G. L. Anderson, et al. (2002). "Risks and benefits of estrogen plus progestin in healthy postmenopausal women: Principal results from the Women's Health Initiative randomized controlled trial." *JAMA* 288 (3): 321–33.

Rotter, N. G. (1988). "Sex differences in the encoding and decoding of

negative facial emotions." *Journal of Nonverbal Behavior,*
12:139–48.

Roussel, S., A. Boissy, et al. (2005). "Gender-specific effects of prenatal stress on emotional reactivity and stress physiology of goat kids." *Horm Behav* 47 (3): 256–66.

Routtenberg, A. (2005). "Estrogen changes wiring of female rat brain during the estrus/menstrual cycle." Society for Neuroscience meeting, Washington, D.C.

Rowe, R., B. Maughan, et al. (2004). "Testosterone, antisocial behavior, and social dominance in boys: Pubertal development and biosocial interaction." *Biol Psychiatry* 55 (5): 546–52.

Rubinow, D., C. Roca, et al. (2002). "Gonadal hormones and behavior in women: Concentrations versus context." In *Hormones, Brain and Behavior,* ed. D. W. Pfaff, vol. 5, 37–74, San Diego: Academic Press.

Rubinow, D. R. (2005). "Reproductive steroids in context." *Arch Women Ment Health* 8 (1): 1–5.

Rubinow, D. R., and P. J. Schmidt (1995). "The neuroendocrinology of menstrual cycle mood disorders." *Ann NY Acad Sci* 771:648–59.

Rubinow, D. R., and P. J. Schmidt (1995). "The treatment of premenstrual syndrome—forward into the past." *N Engl J Med* 332 (23): 1574–75.

Ryan, B. (2000). "Speaking rate, conversational speech acts, interruption, and linguistic complexity." *Clinical Linguistics & Phonetics,* 14 (1): 17–22.

Sa, S. I., and M. D. Madeira (2005). "Neuronal organelles and nuclear pores of hypothalamic ventromedial neurons are sexually dimorphic and change during the estrus cycle in the rat." *Neuroscience* 133 (4): 919–24.

Sabatinelli, D., M. M. Bradley, et al. (2005). "Parallel amygdala and inferotemporal activation reflect emotional intensity and fear relevance." *Neuroimage* 24 (4): 1265–70.

Saenz, C., R. Dominguez, et al. (2005). "Estrogen contributes to structural recovery after a lesion." *Neurosci Lett* 392 (3): 198–201.

Salonia, A., R. E. Nappi, et al. (2005). "Menstrual cycle-related changes in plasma oxytocin are relevant to normal sexual function in healthy women." *Horm Behav* 47 (2): 164–69.

Samter, W. (2002). "How gender and cognitive complexity influence

the provision of emotional support: A study of indirect effects."
*Communication Reports: Special psychological mediators of sex
differences in emotional support* 15 (1): 5–16.

Sanchez-Martin, J. R., E. Fano, et al. (2000). "Relating testosterone
levels and free play social behavior in male and female preschool
children." *Psychoneuroendocrinology* 25 (8): 773–83.

Sandfort, T. G., R. de Graaf, et al. (2003). "Same-sex sexuality and
quality of life: Findings from the Netherlands Mental Health
Survey and Incidence Study." *Arch Sex Behav* 32 (1): 15–22.

Sapolsky, R. M. (1986). "Stress-induced elevation of testosterone
concentration in high ranking baboons: Role of catecholamines."
Endocrinology 118 (4): 1630–35.

Sapolsky, R. M. (2000). "Stress hormones: Good and bad." *Neurobiol
Dis* 7 (5): 540–42.

Sapolsky, R. M., and M. J. Meaney (1986). "Maturation of the
adrenocortical stress response: Neuroendocrine control
mechanisms and the stress hyporesponsive period." *Brain Res* 396
(1): 64–76.

Sastre, J., C. Borras, et al. (2002). "Mitochondrial damage in aging and
apoptosis." *Ann NY Acad Sci* 959:448–51.

Savic, I., H. Berglund, et al. (2001). "Smelling of odorous sex
hormone-like compounds causes sex-differentiated hypothalamic
activations in humans." *Neuron* 31 (4): 661–68.

Sbarra, D. A. (2006). "Predicting the onset of emotional recovery
following nonmarital relationship dissolution: Survival analyses of
sadness and anger." *Pers Soc Psychol Bull* 32 (3): 298–312.

Schirmer, A., and S. A. Kotz (2003). "ERP evidence for a sex-specific
Stroop effect in emotional speech." *J Cogn Neurosci* 15 (8): 1135–48.

Schirmer, A., S. A. Kotz, et al. (2002). "Sex differentiates the role of
emotional prosody during word processing." *Brain Res Cogn Brain
Res* 14 (2): 228–33.

Schirmer, A., S. A. Kotz, et al. (2005). "On the role of attention for the
processing of emotions in speech: Sex differences revisited." *Brain
Res Cogn Brain Res* 24 (3): 442–52.

Schirmer, A., T. Striano, et al. (2005). "Sex differences in the
preattentive processing of vocal emotional expressions."
Neuroreport 16 (6): 635–39.

Schirmer, A., S. Zysset, et al. (2004). "Gender differences in the activation of inferior frontal cortex during emotional speech perception." *Neuroimage* 21 (3): 1114–23.

Schmidt, P. J. (2005). "Depression, the perimenopause, and estrogen therapy." *Ann NY Acad Sci* 1052:27–40.

Schmidt, P. J., N. Haq, et al. (2004). "A longitudinal evaluation of the relationship between reproductive status and mood in perimenopausal women." *Am J Psychiatry* 161 (12): 2238–44.

Schmidt, P. J., J. H. Murphy, et al. (2004). "Stressful life events, personal losses, and perimenopause-related depression." *Arch Women Ment Health* 7 (1): 19–26.

Schmidt, P. J., L. K. Nieman, et al. (1998). "Differential behavioral effects of gonadal steroids in women with and in those without premenstrual syndrome." *N Engl J Med* 338 (4): 209–16.

Schmidt, P. J., L. Nieman, et al. (2000). "Estrogen replacement in perimenopause-related depression: A preliminary report." *Am J Obstet Gynecol* 183 (2): 414–20.

Schmidt, P. J., C. A. Roca, et al. (1998). "Clinical evaluation in studies of perimenopausal women: Position paper." *Psychopharmacol Bull* 34 (3): 309–11.

Schmitt, D. P., and D. M. Buss (1996). "Strategic self-promotion and competitor derogation: Sex and context effects on the perceived effectiveness of mate attraction tactics." *J Pers Soc Psychol* 70 (6): 1185–204.

Schultheiss, O. C., A. Dargel, et al. (2003). "Implicit motives and gonadal steroid hormones: Effects of menstrual cycle phase, oral contraceptive use, and relationship status." *Horm Behav* 43 (2): 293–301.

Schumacher, M. (2002). "Progesterone: Synthesis, metabolism, mechanisms of action, and effects in the nervous system." In *Hormones, Brain and Behavior*, ed. D. W. Pfaff, vol. 3, 683–746. San Diego: Academic Press.

Schutzwohl, A. (2006). "Judging female figures: A new methodological approach to male attractiveness judgments of female waist-to-hip ratio." *Biol Psychol* 71 (2): 223–29.

Schweinsburg, A. D., B. J. Nagel, et al. (2005). "fMRI reveals alteration of spatial working memory networks across adolescence." *J Int Neuropsychol Soc* 11 (5): 631–44.

REFERENCES

Schweinsburg, A. D., B. C. Schweinsburg, et al. (2005). "fMRI response to spatial working memory in adolescents with comorbid marijuana and alcohol use disorders." *Drug Alcohol Depend* 79 (2): 201–10.

Seeman, T. E., B. Singer, et al. (2001). "Gender differences in age-related changes in HPA axis reactivity." *Psychoneuroendocrinology* 26 (3): 225–40.

Seidlitz, L., and E. Diener (1998). "Sex differences in the recall of affective experiences." *J Pers Soc Psychol* 74 (1): 262–71.

Seifritz, E., F. Esposito, et al. (2003). "Differential sex-independent amygdala response to infant crying and laughing in parents versus nonparents." *Biol Psychiatry* 54 (12): 1367–75.

Seurinck, R., G. Vingerhoets, et al. (2004). "Does egocentric mental rotation elicit sex differences?" *Neuroimage* 23 (4): 1440–49.

Shahab, M., C. Mastronardi, et al. (2005). "Increased hypothalamic GPR54 signaling: A potential mechanism for initiation of puberty in primates." *Proc Natl Acad Sci USA* 102 (6): 2129–34.

Sharkin, B. (1993). "Anger and gender: Theory, research and implications." *Journal of Counseling and Development* 71:386–89.

Shaywitz, B. A., S. E. Shaywitz, et al. (1995). "Sex differences in the functional organization of the brain for language." *Nature* 373 (6515): 607–9.

Shaywitz, S. E., F. Naftolin, et al. (2003). "Better oral reading and short-term memory in midlife, postmenopausal women taking estrogen." *Menopause* 10 (5): 420–26.

Shellenbarger, S. (2005). *The Breaking Point: How Female Midlife Crisis Is Transforming today's Women.* New York: Henry Holt.

Sherman, P. W., and B. D. Neff (2003). "Behavioural ecology: Father knows best." *Nature* 425 (6954): 136–37.

Sherry, D. F. (2006). "Neuroecology." *Annu Rev Psychol.* In press.

Sherwin, B. B. (1994). "Estrogenic effects on memory in women." *Ann NY Acad Sci* 743:213–30; discussion 230–31.

Sherwin, B. B. (2005). "Estrogen and memory in women: How can we reconcile the findings?" *Horm Behav* 47 (3): 371–75.

Sherwin, B. B. (2005). "Surgical menopause, estrogen, and cognitive function in women: What do the findings tell us?" *Ann NY Acad Sci* 1052:3–10.

Sherwin, B. B., M. M. Gelfand, et al. (1985). "Androgen enhances sexual motivation in females: A prospective, crossover study of sex steroid administration in the surgical menopause." *Psychosom Med* 47 (4): 339–51.

Shifren, J. L., G. D. Braunstein, et al. (2000). "Transdermal testosterone treatment in women with impaired sexual function after oophorectomy." *N Engl J Med* 343 (10): 682–88.

Shirao, N., Y. Okamoto, et al. (2005). "Gender differences in brain activity toward unpleasant linguistic stimuli concerning interpersonal relationships: An fMRI study." *Eur Arch Psychiatry Clin Neurosci* 255 (5): 327–33.

Shoan-Golan, O. (2004). *Do women cry their own tears? Issues of women's tearfulness, self-other differentiation, subjectivity, empathy and recognition.* Dissertation Abstracts International: Section B: Science and Engineering 65 (1B): 452.

Shors, T. J. (2005). "Estrogen and learning: Strategy over parsimony." *Learn Mem* 12 (2): 84–85.

Shors, T. J. (2006). "Stressful experience and learning across the lifespan." *Annu Rev Psychol.* In press.

Silberstein, S. D., and B. de Lignieres (2000). "Migraine, menopause and hormonal replacement therapy." *Cephalalgia* 20 (3): 214–21.

Silberstein, S. D., and G. R. Merriam (2000). "Physiology of the menstrual cycle." *Cephalalgia* 20 (3): 148–54.

Silk, J. B. (2000). "Ties that bond: The role of kinship in primate societies." In L. Stone, ed., *New Directions in Anthropological Kinship*, 112–21. Boulder, CO: Rowman and Littlefield.

Silk, J. B., S. C. Alberts, et al. (2003). "Social bonds of female baboons enhance infant survival." *Science* 302 (5648): 1231–34.

Silverman, D. K. (2003). "Mommy nearest: Revisiting the idea of infantile symbiosis and its implications for females." *Psychoanalytic Psychology* 20 (2): 261–70.

Silverman, J. (2003). "Gender differences in delay of gratification: A meta analysis." *Sex Roles* 49:451–63.

Simon, R. (2004). "Gender and emotion in the United States." *American Journal of Sociology*, 109:1137–76.

Simon, V. (2005). "Wanted: Women in clinical trials." *Science* 308 (5728): 1517.

References

Singer, E. (2005). "Speech transcript stokes opposition to Harvard head." *Nature* 433 (7028): 790.

Singer, I. (1973). "Fertility and the female orgasm." In *Goals of Human Sexuality*, ed. I. Singer, 159–97. London: Wildwood House.

Singer, T., B. Seymour, et al. (2004). "Empathy for pain involves the affective but not sensory components of pain." *Science* 303 (5661): 1157–62.

Singer, T., and C. Frith (2005). "The painful side of empathy." *Nat Neurosci* 8 (7): 845–46.

Singer, T., B. Seymour, et al. (2006). "Empathic neural responses are modulated by the perceived fairness of others." *Nature* 439 (7075): 466–69.

Singh, D. (1993). "Adaptive significance of female physical attractiveness: Role of waist-to-hip ratio." *J Pers Soc Psychol* 65 (2): 293–307.

Singh, D. (2002). "Female mate value at a glance: Relationship of waisttohip ratio to health, fecundity and attractiveness." *Neuroendocrinology Letters*, 23 (Suppl. 4): 81–91.

Sininger, Y. (1998). "Gender distinctions and lateral asymmetry in the low-level auditory brainstem response of the human neonate." *Hearing Research* 128:58–66.

Skuse, D. (2003). "X-linked genes and the neural basis of social cognition." *Novartis Found Symp* 251:84–98; discussion 98–108, 109–11, 281–97.

Skuse, D., J. Morris, et al. (2003). "The amygdala and development of the social brain." *Ann NY Acad Sci* 1008:91–101.

Slob, A. K., C. M. Bax, et al. (1996). "Sexual arousability and the menstrual cycle." *Psychoneuroendocrinology* 21 (6): 545–58.

Small, D. M., R. J. Zatorre, et al. (2001). "Changes in brain activity related to eating chocolate: From pleasure to aversion." *Brain* 124 (Pt. 9): 1720–33.

Smith, J., M. J. Cunningham, et al. (2005). "Regulation of Kiss1 gene expression in the brain of the female mouse." *Endocrinology* 146 (9): 3686–92.

Smith, M. J., P. J. Schmidt, et al. (2004). "Gonadotropin-releasing hormone-stimulated gonadotropin levels in women with premenstrual dysphoria." *Gynecol Endocrinol* 19 (6): 335–43.

Smith, S. S., and C. S. Woolley (2004). "Cellular and molecular effects of steroid hormones on CNS excitability." *Cleve Clin J Med* 71 (Suppl. 2): S4–10.

Soares, C. N., and O. P. Almeida (2001). "Depression during the perimenopause." *Arch Gen Psychiatry* 58 (3): 306.

Soares, C. N., O. P. Almeida, et al. (2001). "Efficacy of estradiol for the treatment of depressive disorders in perimenopausal women: A double-blind, randomized, placebo-controlled trial." *Arch Gen Psychiatry* 58 (6): 529–34.

Soares, C. N., and L. S. Cohen (2000). "Association between premenstrual syndrome and depression." *J Clin Psychiatry* 61 (9): 677–78.

Soares, C. N., and L. S. Cohen (2001). "The perimenopause, depressive disorders, and hormonal variability." *Sao Paulo Med J* 119 (2): 78–83.

Soares, C. N., L. S. Cohen, et al. (2001). "Characteristics of women with premenstrual dysphoric disorder (PMDD) who did or did not report history of depression: A preliminary report from the Harvard Study of Moods and Cycles." *J Womens Health Gend Based Med* 10 (9): 873–78.

Soares, C. N., H. Joffe, et al. (2004). "Menopause and mood." *Clin Obstet Gynecol* 47 (3): 576–91.

Soares, C. N., J. R. Poitras, et al. (2003). "Effect of reproductive hormones and selective estrogen receptor modulators on mood during menopause." *Drugs Aging* 20 (2): 85–100.

Soares, C. N., J. Prouty, et al. (2005). "Treatment of menopause-related mood disturbances." *CNS Spectr* 10 (6): 489–97.

Sokhi, D. S., M. D. Hunter, et al. (2005). "Male and female voices activate distinct regions in the male brain." *Neuroimage* 27 (3): 572–78.

Soldin, O. P., T. Guo, et al. (2005). "Steroid hormone levels in pregnancy and 1 year postpartum using isotope dilution tandem mass spectrometry." *Fertil Steril* 84 (3): 701–10.

Soldin, O. P., E. G. Hoffman, et al. (2005). "Pediatric reference intervals for FSH, LH, estradiol, T3, free T3, cortisol, and growth hormone on the DPC IMMULITE 1000." *Clin Chim Acta* 355 (1–2): 205–10.

Spelke, E. (2005). "The science of gender and science." *Edge, May* 15.

Spelke, E. S. (2005). "Sex differences in intrinsic aptitude for mathematics and science?: A critical review." *Am Psychol* 60 (9): 950–58.

Speroff, L., P. Kenemans, et al. (2005). "Practical guidelines for post-menopausal hormone therapy." *Maturitas* 51 (1): 4–7.

Speroff, L. (2005). *Clinical Gynecologic Endocrinology and Infertility,* 7th ed. Philadelphia: Lippincott Williams & Wilkins.

Sprecher, S. (2002). "Sexual satisfaction in premarital relationships: Associations with satisfaction, love, commitment, and stability." *J Sex Res* 39 (3): 190–96.

Staley, J. (2006). "Sex differences in diencephalon serotonin transporter availability in major depression." *Biol Psychiatry* 59 (1): 40–47.

Staley, J. K., G. Sanacora, et al. (2006). "Sex differences in diencephalon serotonin transporter availability in major depression." *Biol Psychiatry* 59 (1): 40–47.

Stephen, J. M., D. Ranken, et al. (2006). "Aging changes and gender differences in response to median nerve stimulation measured with MEG." *Clin Neurophysiol.* In press.

Stern, J. M., and S. K. Johnson (1989). "Perioral somatosensory determinants of nursing behavior in Norway rats (*Rattus norvegicus*)." *J Comp Psychol* 103 (3): 269–80.

Stern, J. M., and J. M. Kolunie (1993). "Maternal aggression of rats is impaired by cutaneous anesthesia of the ventral trunk, but not by nipple removal." *Physiol Behav* 54 (5): 861–68.

Stirone, C., S. P. Duckles, et al. (2005). "Estrogen increases mitochondrial efficiency and reduces oxidative stress in cerebral blood vessels." *Mol Pharmacol* 68 (4): 959–65.

Storey, A. E., C. J. Walsh, et al. (2000). "Hormonal correlates of paternal responsiveness in new and expectant fathers." *Evol Hum Behav* 21 (2): 79–95.

Story, L. (2005). "Many women at elite colleges set career path to motherhood." *New York Times,* September 20.

Strauss, J. F., and R. Barbieri (2004). *Yen and Jaffe's Reproductive Endocrinology: Physiology, Pathophysiology, and Clinical Management,* 5th ed. Philadelphia: W. B. Saunders.

Stroud, L. R., G. D. Papandonatos, et al. (2004). "Sex differences in the effects of pubertal development on responses to a corticotropin-releasing hormone challenge: The Pittsburgh psychobiologic studies." *Ann NY Acad Sci* 1021:348–51.

Stroud, L. R., P. Salovey, et al. (2002). "Sex differences in stress responses: Social rejection versus achievement stress." *Biol Psychiatry* 52 (4): 318–27.

Styne, D., D. W. Pfaff, (2002). "Puberty in boys and girls." In *Hormones, Brain and Behavior*, ed. vol. 4, 661–716. San Diego: Academic Press.

Sullivan, E. V., M. Rosenbloom, et al. (2004). "Effects of age and sex on volumes of the thalamus, pons, and cortex." *Neurobiol Aging* 25 (2): 185–92.

Summers, L. (2005). "Conference on Diversifying the Science and Engineering Workforce." NBER transcript, January 14.

Sun, T., C. Pataine, et al. (2005). "Early asymmetry of gene transcription in embryonic human left and right cerebral cortex." *Science* 5729:1794–98.

Sur, M., and J. L. Rubenstein (2005). "Patterning and plasticity of the cerebral cortex." *Science* 310 (5749): 805–10.

Swaab, D. F., W. C. Chung, et al. (2001). "Structural and functional sex differences in the human hypothalamus." *Horm Behav* 40 (2): 93–98.

Swaab, D. F., L. J. Gooren, et al. (1995). "Brain research, gender and sexual orientation." *J Homosex* 28 (3–4): 283–301.

Swerdloff, R., C. Wang, et al. (2002). "Hypothalamic-pituitary-gonadal axis in men." in *Hormones, Brain and Behavior* ed. D. W. Pfaff, vol. v, 1–36. San Diego: Academic Press.

Tanapat, P. (2002). "Adult neurogenesis in the mammalian brain." In *Hormones, Brain and Behavior*, ed. D. W. Pfaff, vol. 3, 779–98. San Diego: Academic Press.

Tang, A. C., M. Nakazawa, et al. (2005). "Effects of long-term estrogen replacement on social investigation and social memory in ovariectomized C57BL/6 mice." *Horm Behav* 47 (3): 350–57.

Tannen, D. (1990). "Gender differences in topical coherence: Creating involvement in best friends' talk." *Discourse Processes: Special gender and conversational interaction* 13 (1): 73–90.

Tannen, D. (1990). *You Just Don't Understand: Women and Men in*

Conversation. New York: William Morrow.

Taylor, S. E., G. C. Gonzaga, et al. (2006). "Relation of oxytocin to psychological stress responses and HPA axis activity in older women." *Psycho Med.* In press.

Taylor, S. E., L. C. Klein, et al. (2000). "Biobehavioral responses to stress in females: Tend-and-befriend, not fight-or-flight." *Psychol Rev* 107 (3): 411–29.

Taylor, S. E., R. L. Repetti, et al. (1997). "Health psychology: What is an unhealthy environment and how does it get under the skin?" *Annu Rev Psychol* 48:411–47.

Tersman, Z., A. Collins, et al. (1991). "Cardiovascular responses to psychological and physiological stressors during the menstrual cycle." *Psychosom Med* 53 (2): 185–97.

Tessitore, A., A. R. Hariri, et al. (2005). "Functional changes in the activity of brain regions underlying emotion processing in the elderly." *Psychiatry Res* 139 (1): 9–18.

Thorne, B. (1983). *Language, Gender and Society.* Boston: Thomson Learning.

Thornhill, R. (1995). "Human female orgasm and mate fluctuating asymmetry." *Animal Behaviour* 50 (6): 1601–15.

Thornhill, R. (1999). "The scent of symmetry: A human sex pheromone that signals fitness?" *Evol Hum Behav* 20:175–201.

Thunberg, M. D. (2000). "Gender differences in facial reactions to fear-relevant stimuli." *Journal of Nonverbal Behavior* 24 (1): 45–51.

Timmers, M. (1998). "Gender differences in motives for regulating emotions." *Pers Soc Psychol Bull* 24:974–86.

Tomaszycki, M. L., H. Gouzoules, et al. (2005). "Sex differences in juvenile rhesus macaque (Macaca mulatta) agonistic screams: Life history differences and effects of prenatal androgens." *Dev Psychobiol* 47 (4): 318–27.

Tooke, W. (1991). "Patterns of deception in intersexual and intrasexual mating strategies." *Ethology and Sociobiology* 12 (5): 345–64.

Toufexis, D. J., C. Davis, et al. (2004). "Progesterone attenuates corticotropin-releasing factor-enhanced but not fear-potentiated startle via the activity of its neuroactive metabolite, allopregnanolone." *J Neurosci* 24 (45): 10280–87.

Tousson, E., and H. Meissl (2004). "Suprachiasmatic nuclei grafts restore the circadian rhythm in the paraventricular nucleus of the hypothalamus." *J Neurosci* 24 (12): 2983–88.

Tranel, D., H. Damasio, et al. (2005). "Does gender play a role in functional asymmetry of ventromedial prefrontal cortex?" *Brain* 128 (Pt. 12): 2872–81.

Trivers, R. (1972). "Parental investment and sexual selection." In *Sexual Selection and the Descent of Man.* ed. B. G. Campbell, 136–79. London: Heinemann Educational.

Tschann, J. M., N. E. Adler, et al. (1994). "Initiation of substance use in early adolescence: The roles of pubertal timing and emotional distress." *Health Psychol* 13 (4): 326–33.

Tuiten, A., G. Panhuysen, et al. (1995). "Stress, serotonergic function, and mood in users of oral contraceptives." *Psychoneuroendocrinology* 20 (3): 323–34.

Turgeon, J. L., D. P. McDonnell, et al. (2004). "Hormone therapy: Physiological complexity belies therapeutic simplicity." *Science* 304 (5675): 1269–73.

Turner, R. A., M. Altemus, et al. (1999). "Preliminary research on plasma oxytocin in normal cycling women: Investigating emotion and interpersonal distress." *Psychiatry* 62 (2): 97–113.

Uddin, L. Q., J. T. Kaplan, et al. (2005). "Self-face recognition activates a frontoparietal 'mirror' network in the right hemisphere: An event-related f MRI study." *Neuroimage* 25 (3): 926–35.

Udry, J. R., and K. Chantala (2004). "Masculinity-femininity guides sexual union formation in adolescents." *Pers Soc Psychol Bull* 30 (1): 44–55.

Udry, J. R., and N. M. Morris (1977). "The distribution of events in the human menstrual cycle." *J Reprod Fertil* 51 (2): 419–25.

Underwood, M. K. (2003). *Social Aggression Among Girls.* New York: Guilford Press.

U. S. Human Resources Services Administration, 2002.

Uvnäs-Moberg, K. (1998). "Antistress pattern induced by oxytocin." *News Physiol Sci* 13:22–25.

Uvnäs-Moberg, K. (1998). "Oxytocin may mediate the benefits of positive social interaction and emotions." *Psychoneuroendocrinology* 23 (8): 819–35.

Uvnäs-Moberg, K. (2003). *The Oxytocin Factor*. New York: Perseus Books.

Uvnäs-Moberg, K., B. Johansson, et al. (2001). "Oxytocin facilitates behavioural, metabolic and physiological adaptations during lactation." *Appl Anim Behav Sci* 72 (3): 225–34.

Uvnäs-Moberg, K., and M. Petersson (2004). "[Oxytocin—biochemical link for human relations: Mediator of antistress, well-being, social interaction, growth, healing . . .]." *Lakartidningen* 101 (35): 2634–39.

Uvnäs-Moberg, K., and M. Petersson (2005). "[Oxytocin, a mediator of antistress, well-being, social interaction, growth and healing]." *Z Psychosom Med Psychother* 51 (1): 57–80.

Uysal, N., K. Tugyan, et al. (2005). "The effects of regular aerobic exercise in adolescent period on hippocampal neuron density, apoptosis and spatial memory." *Neurosci Lett* 383 (3): 241–45.

Van Egeren, L. A. B., S. Marguerite, and M. A. Roach (2001). "Mother–infant responsiveness: Timing, mutual regulation, and interactional context." *Dev Psychol* 37 (5): 684–97.

van Honk, J., A. Tuiten, et al. (2001). "A single administration of testosterone induces cardiac accelerative responses to angry faces in healthy young women." *Behav Neurosci* 115 (1): 238–42.

Vassena, R., R. Dee Schramm, et al. (2005). "Species-dependent expression patterns of DNA methyltransferase genes in mammalian oocytes and preimplantation embryos." *Mol Reprod Dev* 72 (4): 430–36.

Vermeulen, A. (1995). "Dehydroepiandrosterone sulfate and aging." *Ann NY Acad Sci* 774:121–27.

Viau, V. (2006). Personal communication.

Viau, V., B. Bingham, et al. (2005). "Gender and puberty interact on the stress-induced activation of parvocellular neurosecretory neurons and corticotropin-releasing hormone messenger ribonucleic acid expression in the rat." *Endocrinology* 146 (1): 137–46.

Viau, V., and M. J. Meaney (2004). "Testosterone-dependent variations in plasma and intrapituitary corticosteroid binding globulin and stress hypothalamic-pituitary-adrenal activity in the male rat." *J Endocrinol* 181 (2): 223–31.

Vina, J., C. Borras, et al. (2005). "Why females live longer than males: Control of longevity by sex hormones." *Sci Aging Knowledge Environ* 2005 (23): 17.

Vingerhoets, A., and J. Scheir (2000). "Sex Differences in Crying." *Gender and Emotion: Social Psychological Perspectives.* ed. A. H. Fischer, 118–142. New York: Cambridge University Press.

Wager, T. D., and K. N. Ochsner (2005). "Sex differences in the emotional brain." *Neuroreport* 16 (2): 85–87.

Wager, T. D., K. L. Phan, et al. (2003). "Valence, gender, and lateralization of functional brain anatomy in emotion: A meta-analysis of findings from neuroimaging." *Neuroimage* 19 (3): 513–31.

Wagner, H. (1993). "Communication of specific emotions: Gender differences in sending accuracy and communication measures." *Journal of Nonverbal Behavior* 17:29–53.

Walker, C. D., S. Deschamps, et al. (2004). "Mother to infant or infant to mother? Reciprocal regulation of responsiveness to stress in rodents and the implications for humans." *J Psychiatry Neurosci* 29 (5): 364–82.

Walker, Q. D., M. B. Rooney, et al. (2000). "Dopamine release and uptake are greater in female than male rat striatum as measured by fast cyclic voltammetry." *Neuroscience* 95 (4): 1061–70.

Wallen, K. (2005). "Hormonal influences on sexually differentiated behavior in nonhuman primates." *Front Neuroendocrinol* 26 (1): 7–26. Wallen, K. T. (1997). "Hormonal modulation of sexual behavior and affiliation in rhesus monkeys." *Ann NY Acad Sci* 807:185–202.

Wang, A. T., M. Dapretto, et al. (2004). "Neural correlates of facial affect processing in children and adolescents with autism spectrum disorder." *J Am Acad Child Adolesc Psychiatry* 43 (4): 481–90.

Wang, C., D. H. Catlin, et al. (2004). "Testosterone metabolic clearance and production rates determined by stable isotope dilution/tandem mass spectrometry in normal men: Influence of ethnicity and age." *J Clin Endocrinol Metab* 89 (6): 2936–41.

Wang, C., G. Cunningham, et al. (2004). "Longterm testosterone gel (AndroGel) treatment maintains beneficial effects on sexual function and mood, lean and fat mass, and bone mineral density in hypogonadal men." *J Clin Endocrinol Metab* 89 (5): 2085–98.

Wang, C., R. Swerdloff, et al. (2004). "New testosterone buccal system (Striant) delivers physiological testosterone levels: Pharmacokinetics study in hypogonadal men." *J Clin Endocrinol Metab* 89 (8): 3821–29.

Ward, A. M., V. M. Moore, et al. (2004). "Size at birth and cardiovascular responses to psychological stressors: Evidence for prenatal programming in women." *J Hypertens* 22 (12): 2295–301.

Warnock, J. K., S. G. Swanson, et al. (2005). "Combined esterified estrogens and methyltestosterone versus esterified estrogens alone in the treatment of loss of sexual interest in surgically menopausal women." *Menopause* 12 (4): 374–84.

Wassink, T. H., J. Piven, et al. (2004). "Examination of AVPR1a as an autism susceptibility gene." *Mol Psychiatry* 9 (10): 968–72.

Weaver, I. C., N. Cervoni, et al. (2004). "Epigenetic programming by maternal behavior." *Nat Neurosci* 7 (8): 847–54.

Weinberg, M. K. (1999). "Gender differences in emotional expressivity and self-regulation during early infancy." *Dev Psychol* 35 (1): 175–88.

Weiner, C. L., M. Primeau, et al. (2004). "Androgens and mood dysfunction in women: Comparison of women with polycystic ovarian syndrome to healthy controls." *Psychosom Med* 66 (3): 356–62.

Weiss, G., J. H. Skurnick, et al. (2004). "Menopause and hypothalamic-pituitary sensitivity to estrogen." *JAMA* 292 (24): 2991–96.

Weissman, M. M. (2000). "Depression and gender: Implications for primary care." *J Gend Specif Med* 3 (7): 53–57.

Weissman, M. M. (2002). "Juvenile-onset major depression includes childhood- and adolescent-onset depression and may be heterogeneous." *Arch Gen Psychiatry* 59 (3): 223–24.

Weissman, M. M., R. Bland, et al. (1993). "Sex differences in rates of depression: Cross-national perspectives." *J Affect Disord* 29 (2–3): 77–84.

Weissman, M. M., and P. Jensen (2002). "What research suggests for depressed women with children." *J Clin Psychiatry* 63 (7): 641–47.

Weissman, M. M., Y. Neria, et al. (2005). "Gender differences in post-traumatic stress disorder among primary care patients after the

World Trade Center attack of September 11, 2001." *Gend Med* 2 (2): 76–87.

Weissman, M. M., P. Wickramaratne, et al. (2005). "Families at high and low risk for depression: A 3-generation study." *Arch Gen Psychiatry* 62 (1): 29–36.

Weissman, M. M., S. Wolk, et al. (1999). "Depressed adolescents grown up." *JAMA* 281 (18): 1707–13.

Wells, B. E. (2005). "Changes in young people's sexual behavior and attitudes, 1943–1999: A cross-temporal meta-analysis." *Review of General Psychology*, 9 (3): 249–61.

Whitcher, S. J. (1979). "Multidimensional reaction to therapeutic touch in a hospital setting." *J Pers Soc Psychol* 37:87–96.

Williams, N., S. L. Williams, et al. (1997). "Mild metabolic stress potentiates the suppressive effect of psychological stress on reproductive function in female cynomolgus monkeys." Endocrine Society meeting, Minneapolis, abstract PI367.

Wilson, B. C., M. G. Terenzi, et al. (2005). "Differential excitatory responses to oxytocin in sub-divisions of the bed nuclei of the stria terminalis." *Neuropeptides* 39 (4): 403–7.

Wilson, M. E., A. Legendre, et al. (2005). "Gonadal steroid modulation of the limbic-hypothalamic-pituitary-adrenal (LHPA) axis is influenced by social status in female rhesus monkeys." *Endocrine* 26 (2): 89–97.

Windle, R. J., Y. M. Kershaw, et al. (2004). "Oxytocin attenuates stress-induced c-fos mRNA expression in specific forebrain regions associated with modulation of hypothalamo-pituitary-adrenal activity." *J Neurosci* 24 (12): 2974–82.

Winfrey, O. (2005). "Turning fifty." *Oprah*, May.

Wise, P. (2003). "Estradiol exerts neuroprotective actions against ischemic brain injury: Insights derived from animal models." *Endocrine* 21 (1): 11–15.

Wise, P. (2006). "Estrogen therapy: Does it help or hurt the adult and aging brain? Insights derived from animal models." *Neuroscience*. In press.

Wise, P. M. (2003). "Creating new neurons in old brains." *Sci Aging Knowledge Environ* (22): PE13.

Wise, P. M., D. B. Dubal, et al. (2005). "Are estrogens protective or risk factors in brain injury and neurodegeneration? Reevaluation after the Women's Health Initiative." *Endocr Rev* 26 (3): 308–12.

Witelson, S. F., H. Beresh, et al. (2006). "Intelligence and brain size in 100 postmortem brains: Sex, lateralization and age factors." *Brain* 129 (Pt. 2): 386–98.

Witelson, S. F. (1995). "Women have greater density of neurons in posterior temporal cortex." *J Neurosci* 15 (5, Pt. 1): 3418–28.

Wood, G. E., and T. J. Shors (1998). "Stress facilitates classical conditioning in males, but impairs classical conditioning in females through activational effects of ovarian hormones." *Proc Natl Acad Sci USA* 95 (7): 4066–71.

Woods, N. F., E. S. Mitchell, et al. (2000). "Memory functioning among midlife women: Observations from the Seattle Midlife Women's Health Study." *Menopause* 7 (4): 257–65.

Woolley, C. a. R. C. (2002). "Sex steroids and neuronal growth in adulthood." In *Hormones, Brain and Behavior*, ed. D. W. Pfaff, vol. 4, 717–78.

Woolley, C. S., H. J. Wenzel, et al. (1996). "Estradiol increases the frequency of multiple synapse boutons in the hippocampal CA1 region of the adult female rat." *J Comp Neurol* 373 (1): 108–17.

Wrangham, R. W. (1980). "An ecological model of female-bonded primate groups." *Behaviour* 75:262–300.

Wrangham, R. W., and B. B. Smuts (1980). "Sex differences in the behavioural ecology of chimpanzees in the Gombe National Park, Tanzania." *J Reprod Fertil Suppl*, Suppl. 28: 13–31.

Wrase, J., S. Klein, et al. (2003). "Gender differences in the processing of standardized emotional visual stimuli in humans: A functional magnetic resonance imaging study." *Neurosci Lett* 348 (1): 41–45.

Wright, J., F. Naftolin, et al. (2004). "Guidelines for the hormone treatment of women in the menopausal transition and beyond: Position statement by the Executive Committee of the International Menopause Society." *Maturitas* 48 (1): 27–31.

Xerri, C., J. M. Stern, et al. (1994). "Alterations of the cortical representation of the rat ventrum induced by nursing behavior." *J Neurosci* 14 (3, Pt. 2): 1710–21.

Yamamoto, Y., C. S. Carter, et al. (2006). "Neonatal manipulation of oxytocin affects expression of estrogen receptor alpha."

Neuroscience 137 (1): 157–64.

Yamamoto, Y., B. S. Cushing, et al. (2004). "Neonatal manipulations of oxytocin alter expression of oxytocin and vasopressin immunoreactive cells in the paraventricular nucleus of the hypothalamus in a gender-specific manner." *Neuroscience* 125 (4): 947–55.

Yen, S., R. Jaffe (1991). *Reproductive endocrinology: Physiology, pathophysiology, and clinical management.* Philadelphia: W. B. Saunders.

Yonezawa, T., K. Mogi, et al. (2005). "Modulation of growth hormone pulsatility by sex steroids in female goats." *Endocrinology* 146 (6): 2736–43.

Young, E., C. S. Carter, et al. (2005). "Neonatal manipulation of oxytocin alters oxytocin levels in the pituitary of adult rats." *Horm Metab Res* 37 (7): 397–401.

Young, E. A., H. Akil, et al. (1995). "Evidence against changes in corticotroph CRF receptors in depressed patients." *Biol Psychiatry* 37 (6): 355–63.

Young, E. A., and M. Altemus (2004). "Puberty, ovarian steroids, and stress." *Ann NY Acad Sci* 1021:124–33.

Young, E. A. (2006). Personal communication.

Young, E. A. (2002). "Stress and anxiety disorders." In *Hormones, Brain and Behavior*, ed. D. W. Pfaff, vol. 5, 443–66. San Diego: Academic Press.

Young, L. J., M. M. Lim, et al. (2001). "Cellular mechanisms of social attachment." *Horm Behav* 40 (2): 133–38.

Yue, X., M. Lu, et al. (2005). "Brain estrogen deficiency accelerates A [beta] plaque formation in an Alzheimer's disease animal model." *Proc Natl Acad Sci USA* 102 (52): 19198–203.

Zahn-Waxler, C., B. KlimesDougan, et al. (2000). "Internalizing problems of childhood and adolescence: Prospects, pitfalls, and progress in understanding the development of anxiety and depression." *Dev Psychopathol* 12 (3): 443–66.

Zahn-Waxler, C., M. RadkeYarrow, et al. (1992). "Development of concern for others." *Dev Psychol* 28:126–36.

Zak, P. J., R. Kurzban, et al. (2005). "Oxytocin is associated with human trustworthiness." *Horm Behav* 48 (5): 522–27.

Zald, D. H. (2003). "The human amygdala and the emotional

evaluation of sensory stimuli." *Brain Res Brain Res Rev* 41 (1):
88–123.

Zemlyak, I., S. Brooke, et al. (2005). "Estrogenic protection against
gp120 neurotoxicity: Role of microglia." *Brain Res* 1046 (1–2):
130–36.

Zhang, T. Y., P. Chretien, et al. (2005). "Influence of naturally
occurring variations in maternal care on prepulse inhibition of
acoustic startle and the medial prefrontal cortical dopamine
response to stress in adult rats." *J Neurosci* 25 (6): 1493–502.

Zhou, J., D. W. Pfaff, et al. (2005). "Sex differences in estrogenic
regulation of neuronal activity in neonatal cultures of ventromedial
nucleus of the hypothalamus." *Proc Natl Acad Sci USA* 102 (41):
14907–12.

Zimmerberg, B., and E. W. Kajunski (2004). "Sexually dimorphic
effects of postnatal allopregnanolone on the development of
anxiety behavior after early deprivation." *Pharmacol Biochem Behav*
78 (3): 465–71.

Zonana, J., and J. M. Gorman (2005). "The neurobiology of
postpartum depression." *CNS Spectr* 10 (10): 792–99, 805.

Zubenko, G. S., H. B. Hughes, et al. (2002). "Genetic linkage of region
containing the CREB1 gene to depressive disorders in women
from families with recurrent, early-onset, major depression." *Am J
Med Genet* 114 (8): 980–87.

Zubieta, J. K., T. A. Ketter, et al. (2003). "Regulation of human
affective responses by anterior cingulate and limbic muopioid
neurotransmission." *Arch Gen Psychiatry* 60 (11): 1145–53.

INDEX